Introduction to Sheep Farming

R. G. Johnston

GRANADA
London Toronto Sydney New York

Granada Publishing Limited — Technical Books Division
Frogmore, St Albans, Herts AL2 2NF
and
36 Golden Square, London W1R 4AH
866 United Nations Plaza, New York, NY 10017, USA
117 York Street, Sydney, NSW 2000, Australia
100 Skyway Avenue, Rexdale, Ontario, Canada M9W 3A6
61 Beach Road, Auckland, New Zealand

Copyright © R. G. Johnston 1983

British Library Cataloguing in Publication Data
Johnston, R. G.
Introduction to sheep farming.
1. Sheep
636.3   SF375

ISBN 0-246-11962-4

First published in Great Britain 1983 by Granada Publishing Ltd

Printed in Great Britain by Mackays of Chatham, Kent

Granada ®
Granada Publishing ®

Contents

Foreword

by J. L. Read, Head of Sheep Improvement Services, Meat and Livestock Commission

Sheep production is the most widespread form of animal husbandry. It encompasses a broad range of methods and scale of operation and a diversity of end products, ranging from carcase meat through wool and skins to milk and dairy products. In many situations sheep production has come to be regarded as an extractive form of husbandry utilising natural low quality grazing which is unsuitable for any other form of livestock production. At the opposite extreme there are intensive systems of management on temperate grassland with high stocking rates, prolific ewes and the use of fertiliser and supplementary feed to give outputs of a tonne per hectare of lamb.

Recognising the problems that the complexity of the industry and methods of production present to the student and enquiring farmer, the author of this book has set out to provide a guide to the principles that underlie all types of sheep production. This is a sound, if demanding, approach designed to make the reader think through the production implications of the biology of the sheep and the environment and the degree to which each can be economically modified. It does not seek to be a cook book of instant recipes for success with sheep; the author is too experienced and sagacious to fall into that trap. Readers will have cause to thank him for giving them instead the raw material of knowledge of the basic elements and an understanding of how they interact in defined production systems.

In a world with a large unsatisfied demand for animal protein and consumers with an increasing aversion to the products of factory farming, the future for sheep meat, the natural product, seems to be assured. But the demand will remain unsatisfied unless the sheep producer is able to operate profitably. Knowledge and application are the key to success in this as in every other farming enterprise and while all knowledge cannot be won from books, as the author is careful to point out, there is a great body of knowledge on sheep to which this work provides a key.

Preface

The primary purpose of this book is to provide a text for agricultural students and for those farmers wishing to set up a sheep enterprise.

It is the usual practice for colleges and universities to insist that agricultural students have a minimum of one year's practical farming experience before entering an academic course. In the days when mixed farming was common practice this ensured that most students had at least a passing acquaintance with most of the crops and stock common to their area. Nowadays, however, the degree of specialisation which prevails in most developed countries makes it very difficult for the student to become familiar with a range of domestic animals in a short period of time. The modern teacher, therefore, can no longer assume that his students have even an elementary knowledge of a particular species of animal. This point has been borne in mind while writing this book and little or no previous knowledge of sheep is assumed on the part of the reader.

The book is based on my own teaching and is arranged in a sequence suitable for forming the basis of any general lecture course. It is primarily directed towards students within the group from National Certificate in Agriculture to Ordinary and Higher National Diploma level. It is also hoped that it may prove helpful to practical sheep farmers in outlining the science behind their craft. I have spent a lifetime working with sheep and lecturing on livestock. This, combined with a wide reading and discussions with experts from both the academic and practical side of sheep husbandry, hopefully has resulted in a book without too many sins of commission or omission.

There are many excellent books on various aspects of sheep husbandry and on these I have drawn. A list for further reading is incorporated at the end of the book with indications of the particular strengths of each book.

Of the many friends with whom I have discussed sheep in general and this book in particular, from the academic side I should like to thank Mr J. L. Read and Mr J. Stephenson for reading and commenting on the script, and Mr Read for kindly writing the foreword. I also have to thank Professor J. M. M. Cunningham, Dr Clive Dalton and Mr P. J. Davies for their helpful advice. From the field of practical farming and farm management I am indebted to Mr Bruce Blacklock, Mr T. D. Borland, Mr T. A. Dykes, Mr R. W. Emery, Mr E. Hughes and Mr A. B. Mason.

My thanks also go to the Meat and Livestock Commission for allowing me to use diagrams and tables from their own publications, particularly *Feeding the Ewe* and *Commercial Sheep Production Year Book* and also to H.M.S.O. for similar help. I also thank Dr Clive Dalton for kindly permitting me to use Figs. 6.1, 6.2, 6.3, 7.8, and 13.4, and Dr Tony Kempster of the Meat and Livestock Commission for kindly making the M.L.C. photographs on carcase quality available. Dr Kempster also supplied Fig. 7.2. Mr A. W. Bryson of the Blackface Sheep Breeders' Association kindly supplied Fig. 1.3, and Mr J. A. Minto of Biggar supplied Fig. 9.1. Thanks are also due to Farmers' Weekly for kindly providing Figs. 1.1, 1.2 and 1.4.

I wish to thank Mr Richard Miles of Granada Publishing for much helpful advice and necessary prodding without which the book would not have come to pass.

Finally grateful thanks are due to the people who did the hard work, my daughter-in-law Liz who helped with the script and Mrs N. Elliott who so gallantly typed it.

R. G. Johnston

1 Introduction

Origins

Let us begin with the place of sheep in the zoological system. Sheep belong to the order of *Ungulates* or hoofed mammals, and to the suborder *Pecora* or true ruminants. Their family is the *Bovidae* which are hollow-horned and the genus is *Ovis*. All wild sheep such as the Mouflon, Barbary Sheep and the Rocky Mountain Big Horns belong to this genus. Sheep fossils are known from the rocks of some twenty million years ago, but their geological history is tenuous and those who are familiar with the beautiful record of the rocks depicting the evolution of the horse from Eohippus to the modern Shire will seek in vain for a similar story of the sheep. The majority, if not all, of modern wild sheep are found in rugged, rocky, mountainous terrain, and it appears that their ancestors took to the hills at an early period in their development.

At the present time sheep are kept in many parts of the world under a wide range of environmental conditions. They are primarily animals of the high ground and semi-arid parts of the world and are rarely found in marshy or heavily wooded country. Modern breeds of western European origin, for instance, are found in their greatest concentration in places such as the high ground of Britain, Spain and New Zealand, the Karoo of South Africa and the drier farming areas of Australia and South America although there are, of course, exceptions to this pattern such as Romney Marsh in Britain and low-lying areas of New Zealand like the Canterbury Plain.

The physical make-up of the sheep

The sheep, being a mammal, is a warm-blooded animal which maintains its normal body temperature at about 40°C. The resting pulse rate is between 75 and 80 beats per minute, with a respiration rate of 20 to 30 for the same period. Table 1.1 sets out some of the basic physical and physiological properties of sheep.

Table 1.1 Basic physical and physiological properties of sheep.

Average temperature	40°C
Average pulse rate	75-80 per minute
Average respiration rate	20-30 per minute
Oestrus cycle	16 days
Gestation period	147 days
Litter size	Normally 1-3, exceptionally up to 7
Age of sexual maturity	
Rams	7 months
Ewes	7 months
Natural lifetime	8-10 years

Sheep vary greatly in size between one breed and another and also within breeds. The adult males are substantially larger than females of the same breed. Weight variations are from about 30 kg for a Welsh Mountain ewe off hard ground to 100 kg for a longwool ewe such as a Lincoln. Body weight variations within breeds can also be quite marked and mainly reflect differences in nutrition. It is particularly marked in the mountain breeds such as the Scottish Black-face, where the difference in size between sheep from really high hills and those from areas verging on the margin of cultivatable land can be quite substantial. Dentition, digestion and related bodily factors will be covered in the next chapter.

The wool-growing characteristics of different breeds are as variable as their size, both for quantity and quality. The Wiltshire Horn has virtually no wool, and that which it has is regularly being cast. Merinos have relatively heavy fleeces of very fine wool. The western European white faced breeds of sheep tend to have long lustrous wools which are not particularly fine. The majority of British moun-tain breeds with dark faces, such as the Scottish Blackface, Swaledale and Rough Fell, have long coats of coarse wool which often includes a considerable intermingling of kemp (a very coarse type of fibre which is also hollow). These last wools are classed as carpet wools as are the fleeces of most North African and Asiatic sheep. The wools of other breeds and crosses are classed by the wool trade as 'cross-bred'.

The breeding cycle of the sheep

Turning to what may be termed the life style of the sheep, the first point of interest is the time taken to reach maturity which, for all practical purposes, means sexual maturity. This can be reached at 7 months of age but nutrition is important: underfeeding will lengthen the time taken to come to breeding condition. The majority of *low ground* females which have been adequately fed, and born not too late in the season will take the ram in their first autumn to lamb the following spring. The lamb crop from these young sheep is normally in the region of 70%, i.e. 70 lambs weaned per 100 ewes mated. Under conditions of restricted nutrition such as on hill farms, however, sheep are not normally put to the ram until they are 18 months of age or older. Well-nourished rams will work from the age of about 7 months, but the number of ewes allocated to the ram lambs must always be substantially fewer than those presented to adult rams under the same conditions.

The onset of heat in a ewe is controlled mainly by day length, sexual activity being induced by a shortening of the day length. This effect is triggered when the period of daylight changes from above 13-14 hours to less. Conversely the mating season ends when the day length rises above 13-14 hours. Here again, some variations are met with, not only between breeds, but between individuals within breeds. This reaction is controlled by the action of light via the eye on the anterior pituitary gland (a small ductless gland at the base of the brain).

When the ewe lamb becomes sexually mature the breeding cycle which ensues is as follows:

- Oestrus takes place and this can vary in duration from a few hours to 3 days, but the average duration is about 27 hours. Unlike the sow or the cow, ewes show less obvious signs of being in season and the only reliable way to detect heats is to run a ram with the flock.
- If on mating the sheep does not become pregnant she will normally come into oestrus again in about 16 days. The period is variable and can be as little as 7 days or as long as 60 days, but these are very extreme cases. The normal fluctuations vary between 14 and 20 days, the average being 16.
- Most ewes exhibit *lactation anoestrus*, i.e. they do not come on heat when secreting milk and, additionally, most ewes of western European origin, i.e. sheep from relatively high latitudes, will take the ram in only autumn and winter. A well-known exception to this rule is the Dorset Horn and its derivative the Polled Dorset: Merinos tend to show similar behaviour, having an extended breeding season themselves.

Rams appear to remain fertile throughout the year although during the period the ewes are quiescent they exhibit little sexual urge. Having become pregnant the gestation period of the ewe lasts about 147 days, or 5 months. Ewes are remarkably constant in respect of lambing and do not show the same vagaries as cows and mares. The size of the litter to which a ewe gives birth varies with a number of factors. The breed of the animal is important. Hill breeds and those kept in arid areas tend to have only one lamb, while a low ground breed such as the Finnish Landrace — the most prolific breed in Europe — has been known to give birth to seven. In an extensive farming situation, one lamb is all that is desirable for each ewe, but in an intensive meat lamb producing unit the number of lambs sold per ewe is of paramount importance and twins are desirable.

Factors affecting the breeding performance of ewes

The genetic factor is naturally of prime importance, and some breeds are much more prolific than others: compare, for instance, the Finnish Landrace with its large litter and a Merino with a single lamb. Nevertheless the heritability of prolificacy is low and selection within a breed for multiple births is a slow process. The simplest way of raising the prolificacy of a flock is to introduce genetic material from a breed that is itself highly prolific. Introducing Finnish Landrace 'blood' into British and the Australian Booroola into New Zealand sheep stocks to raise fertility are good examples.

Having regard to the individual ewe within a flock, the following factors affect the likelihood of multiple births:

- On average there is a rise with age in the number of lambs born per ewe, until the age of about 3 years where a plateau develops and then there tends to be a decline after about five years.
- In any particular breed there is a tendency for the larger, heavier ewes to breed more twins and triplets than their smaller sisters.
- The time of mating also influences the number of multiple births, as ewes lambed very early in the season tend to give smaller crops than those lambed in March and April under western European conditions. Most twins and triplets are born from shortly after lambing commences until the middle of the lambing period.
- A factor that becomes evident from consulting lambing records is *repeatability*. Ewes which have produced big litters in the past tend to continue so doing until the litter size decreases with old age. Prolificacy varies markedly between different ewes of the

same age and breed in the same flocks and treated the same. This means that a simple method of raising the lambing performance of a flock is to cull after two or three pregnancies those ewes which do not produce more than one lamb a time.

- *Flushing* is the low ground sheep farming practice of increasing fertility by putting the ewes on to a higher plane of nutrition before introducing them to the rams. It is achieved by such means as providing them with a hay aftermath, reseeded grass, or other succulent crop, or by feeding concentrates. Flushing is started 14-21 days before the rams are turned in to the ewes. In the case of lean ewes this practice leads to increased ovulation. This in turn has given rise to the idea amongst many farmers that ewes should be brought down in condition during the dry period prior to flushing but this bringing down in condition is not necessary: ewes which go to the ram in a well-fleshed condition will breed perfectly satisfactorily. This can be seen by observing the performance of meat lamb producing flocks where the ewes have got back into good condition long before tupping time and yet perform very adequately at the next lambing. The modern practice is to try to bring ewes back to a satisfactory body weight in 6-8 weeks.

This concentration on nutrition should not be taken to imply that it does not matter whether or not the ewes are over-fed. The watchword for ewes, as for all breeding stock, is that they should be fit but not fat. The encouragement of multiple ovulation in extensively kept sheep, as on hill farms, is not usually a sensible operation and this point will be discussed later on.

Milk production in the ewe

Satisfactory milk production is vital. A ewe with a large litter but a low milk yield is not the best foundation for success! The milk is secreted in an udder which is divided into two sections, each section having one teat. As with many other mammals, supernumerary teats are not uncommon.

Milk yield depends largely on heredity, assuming a proper level of nutrition. Some breeds such as the East Friesland milk sheep will give as much as 5-600 litres in a 200-day lactation whereas the average yield of the French Lacaune breed from which Roquefort cheese is made is 100-150 litres in a six month lactation. Some ewes, of course, will give very much higher yields. In areas where sheep are kept for such purposes some flocks are machine-milked. Of the sheep breeds used basically for meat production longwool breeds such as the Border Leicester and their crosses give good yields. Most British hill breeds give good yields under conditions of adequate feeding. The yields of the cross breeds, such as Halfbreeds, Greyfaces and the like used in meat lamb production appear to be in the region of 7 to 18 litres per week.

As with the cow the composition of ewe milk is variable, but it should be noted that on average it has a much higher solid content than cow milk and is also richer than that of the goat (see Table 1.2). It is particularly high in fat, being about double that of cows other than Channel Islanders.

Table 1.2 Average composition by weight of ewe milk compared with cow and goat milk.

	Ewe	Cow	Goat
Fat (%)	7.4	3.7	4.6
Protein (%)	5.8	3.4	4.4
Lactose (%)	4.8	4.8	4.2
Minerals (%)	1.0	0.75	0.8

It is important that the high compositional quality of milk is borne in mind when feeding the lactating ewe, and even more so when providing milk substitutes for artificially reared lambs. Assuming that the ewe is inherently capable of good milk yield, she will be unable to attain this unless conditions are favourable. The body condition of the ewe at lambing is of major importance and is the foundation for a good lactation yield.

The ewe, like the cow, needs 'steaming up' if she is to give a good yield. Unless she is well-fed her udder will not be properly developed and unless she has substantial body reserves she will not be able to 'milk off her back' and give a lot of milk in early lactation. The ewe also needs to be well-fed after lambing, particularly for protein, or her milk yield will not be sustained regardless of initial body condition. Every effort should be made to see that the ewe is not held back in the early period after lambing by being kept short of water, running a high temperature, being underfed or inappropriately fed, or being subjected to harassment. Similarities exist between the ewe and the cow and it would seem reasonable to suppose that they react similarly to bad conditions in early lactation. It is well known that if a cow suffers from adverse factors in early lactation she will not reach her potential peak yield, neither will she achieve her lactation potential. Observation of lambs whose mothers have received set-backs in early lactation support this view. They fall behind their more favoured contemporaries and never regain the lost ground in a reasonable time. This is of great importance in meat lamb production where missing a particular market could be heavily penalised by a fall in selling price.

Factors affecting the birth weight and growth of lambs

The *breed* of the ewe has a major influence in the size of the lamb: the larger breeds tend to have the larger lambs. This is also true within breeds. The larger ewes tend to have larger lambs than the smaller. Also the lambs of the older ewes tend to be larger than those of their younger sisters.

The size of the lamb at birth is affected mainly by the mother but the *sire* does have an influence: the offspring of large sires tend to be larger than those of smaller animals. The *conformation* of the sire is also important. When large rams are used on small ewes, big-headed, square-shouldered rams tend to increase the number of *dystocias* or difficult lambings. The chief influence of the ram of importance in meat lamb production, however, is that of *growth rate* and *carcase quality*.

Heavier lambs are stronger and more viable than lighter animals, and the birth weight is affected considerably by the *number of lambs born per litter*. As one would expect, the larger the number of offspring the lower the individual birth weights. The figures given below are from the flock of Mule ewes crossed with a Suffolk ram, and will serve as an indication of the sort of differences that do occur. Of 66 ewes lambing, 16 had triplets, 38 twins, and 12 singles. The average lamb weights were: triplets 3.57 kg; twins 4.58 kg; singles 5.54 kg. This shows the triplets to be about 22% lighter than the twins, whereas the singles are 17% heavier. In Britain average national figures are available for different breeds and crosses.

The *sex* of the lamb also influences birth weight. In typical lowland meat lamb producing flocks the males are, on average, 0.25 kg to 0.5 kg heavier than the females.

Another important factor affecting the birth weight is the *nutritional status of the mother*. The last 6 to 8 weeks of pregnancy are of paramount importance, as it is during this period that the foetus grows at its maximum rate. During the first 3 months it makes little demand on the ewe. The problems of nutrition and birth weights have been very comprehensively explored by experimenters. Birth weight matters crucially because not only are the lighter lambs less viable than the heavier, but they also grow more slowly.

Under-feeding pregnant ewes in late pregnancy produces lambs which are below potential optimum weight at birth and are, therefore, less resistant to exposure and disease. They will also grow more slowly. The underfed ewes will also give less milk, further slowing down the already checked growth rate. It must be noted, however, that what one is looking for is *not* the largest lamb possible. Fat ewes carrying very large lambs is a sure recipe for a disastrous lambing both from dystocias and quite possibly from pregnancy toxaemia. On the other hand seriously undernourished ewes are also greatly at risk from the same complaint. The lesson to be drawn from this situation is the same as for so many in animal production: success depends on striking a proper *balance*.

Having been born with satisfactory body weight the lamb's first vital requirement is its mother's first milk or *colostrum*. By virtue of its antibody content colostrum gives protection against diseases with which the dam has been in contact or vaccinated against. The continued progress of the lamb will depend on a full milk supply, especially during the first three or four weeks of life, i.e. before it starts eating a significant quantity of solid food.

Milk supply is of great importance as where twins are born and one dies the survivor tends to grow at a rate similar to that of a lamb born a single. Lambs of different breeds and crosses grow at different rates but in early life this difference seems to reflect mainly the milking capacity of the ewes. Notwithstanding what has already been said in the preceeding paragraphs one must take note of the remarkable ability shown by the ewe to nourish the foetus at the expense of her own body tissues and having lambed, to 'milk off her back' to her own physical detriment. This phenomenon is to be seen year by year in hill flocks, but in the years of extreme deprivation many ewes are defeated in the unequal battle and not only are their lambs lost but many ewes as well. The subject of supplementary feeding in marginal conditions will be discussed later.

Little reference has so far been made to rams as it is more appropriate to discuss the ram later when considering the selection of males for a particular husbandry system.

Sheep breeds

SHEEP BREEDS IN BRITAIN

There are about 31 million sheep at present in Britain, of which about 12 million are ewes and 3 million are shearlings. The average flock size is 175 but about 40% of breeding sheep are in flocks of 500 or over. There are 50 recognised breeds at present, plus a number of local types and a variety of cross-breeds, although most other countries manage with far fewer breeds. New Zealand, for instance, relies on two basic breeds, the Merino on high ground and the Romney on lower ground. These and their crosses mated to Down rams, usually the Southdown, form the basis of their meat lamb production.

British breeds are usually classified as longwools, shortwools, and mountain, moorland and hill breeds.

The *longwools* are large, hornless, with heavy fleeces and white faces and legs. They are suited to the more fertile areas where food is plentiful and living is fairly easy. Their numbers have declined substantially over the past century and their importance now rests on the production of rams for breeding meat lamb mothers (see later in Chapter 4 under *Stratification*). The *shortwools* are smaller animals which mature early, have dark faces and legs, are hornless, and have a fine, crimped short wool. The wool is creamy in colour and sometimes mixed with brown or black. Mutton conformation and quality are excellent and these breeds are adapted to areas with drier, lighter soils. *Mountain, moorland and hill breeds* are small, active and hardy with good mothering ability, and are particularly suited to rigorous climatic conditions where soil is poor and food is scarce.

Table 1.3 sets out some of the main breeds according to category.

SHEEP BREEDS ELSEWHERE

In semi-arid and upland areas elsewhere in the world, in Australasia, South Africa and the Americas, the premier breeds have been the Merino and its derivatives such as the Rambouillet. Such sheep are crossed with English longwools to produce meat lamb mothers which, when crossed with Down rams, form the basis of the sheep meat production industries in these areas.

Names of sheep according to sex and stage of life

The number of regional names given to sheep is both remarkable and confusing. In no other species of farm animal is such a great diversity

Fig. 1.1 English Leicester ram (*Farmers' Weekly*).

Table 1.3 Some modern British breeds.

Longwools	Shortwools	Mountain, moorland and hill breeds
English Leicester	Down breeds:	Scottish Blackface
Lincoln	Southdown	South Country Cheviot
Border Leicester	Hampshire	(or Cheviot)
Bluefaced Leicester	Dorset Down	Swaledale
Kent or Romney Marsh	Suffolk	Welsh Mountain
Devon and Cornwall	Shropshire	Exmoor Horn
longwool	Oxford Down	Clun Forest
Wensleydale	Horned whitefaced	Kerry Hill
	breeds:	Rough Fell
Teeswater	Dorset Horn	Dales Bred

Longwools	Shortwools	Mountain, moorland and hill breeds
	Wiltshire Horn Polled whitefaced breeds: Ryeland Polled Dorset	Dartmoor Herdwick

Note: The North Country Cheviot, the Clun Forest and the Kerry Hill are normally listed as hill sheep but they are prolific, milky ewes which are quite fast-growing. Consequently, they are frequently used as meat lamb mothers on low ground farms.

Fig. 1.2 Suffolk ewe (*Farmers' Weekly. Photograph: Douglas Low*).

of names applied to animals of different sexes and ages. In the British Isles these names vary not only from country to country but also from county to county. Table 1.4 summarises the names which are most generally accepted.

Fig. 1.3 Scottish Blackface (*Blackface Sheep Breeders' Association*).

Fig. 1.4 A 2½-year-old merino ram (*Farmers' Weekly. Photograph: L T Sardone*).

Table 1.4 Names applied to sheep according to sex and stage of life.

Age	Female	Male (entire)	Male (castrated)	Notes
Birth to weaning (0 to 14–16 weeks old)	Ewe lamb	Ram lamb / Tup lamb	Wether lamb	A sheep is called a *lamb* until weaned
Weaning to shearing (14–16 weeks to about 15 months old)	Ewe hogg or hogget	Ram hogg or hogget / Tup hogg or hogget	Wether hogg or wether hogget	Wool of first shearing is called *hogget* wool
First to second shearing	Shearling ewe / Gimmer / Two-toothed ewe / Theave (in South of England)	Shearling ram / Shearling tup / Two-toothed ram	Shearling wether / Two-toothed wether	Once a female has lambed she is classed as a *ewe*
Second to third shearing	Three-shear ewe / Six-toothed ewe	Two-shear ram / Two-shear tup	Wethers are no longer kept at this age	
Third to fourth shearing	Three-shear ewe / Six-toothed ewe	Three-shear ram / Three-shear tup	—	
After fourth shearing	Four-shear ewe / Full-mouthed ewe	Full-mouthed ram / Full-mouthed tup	—	Sheep after the full-mouthed stage are classed as *aged*

Points to remember

1. Physical and physiological properties of sheep:

Average temperature:	40°C.
Average pulse rate:	75–80 per minute.
Average respiration rate:	20–30 per minute.
Oestrus cycle:	16 days.
Gestation period:	147 days.

2. Factors affecting the birth weight of lambs and their subsequent growth:

   Breed and size of ewe.
   Breed and size of sire (though to a lesser extent).
   Nutrition of ewe.
   Sex of lamb.
   Number of lambs born.

3. Factors affecting multiple births:

   Age of dam.
   Time of mating.
   Body weight and condition of ewe.
   Repeatability and genetic effect.

4. Factors affecting milk yield:

   Heredity.
   Body condition and nutrition.

2 Nutrition and feeding

The sheep in the wild is a grazing animal sustaining itself on grass, legumes and similar herbage. It also does some browsing off shrubs and trees but not to the same extent as its cousin the goat. It is a true *ruminant*, having the following advantages over simple-stomached animals:

- It can digest cellulose and hemicellulose.
- It can make use of non-protein nitrogen.
- It can synthesise some of the vitamins which it requires.

This means that it can live off relatively inexpensive foods, that it can be fed without recourse to concentrates and it does not compete for grain and concentrated protein foods with human beings or simple-stomached animals such as swine, or with poultry. This statement is not intended to imply that concentrates should not be fed to sheep; indeed, concentrates are most important feeds in many intensive sheep systems.

The normal food of domesticated sheep is pasture herbage, comprising grasses, clovers and various herbs. In mountainous and semi-arid areas low shrubs are also eaten. Typical examples of these are species of *Calluna* (heather) and *Vaccinium* and, in Australia, *Atriplex*.

Where sheep are kept intensively, arable crops may be grown for their special use and herbage conserved for supplementary feeding. In areas of reasonable rainfall the fodder crops are mainly succulent in nature – swedes, turnips, rape, kale and similar crops, while in drier areas the crop might be irrigated lucerne (alfalfa). The conserved materials are in the main hay and silage.

Before commencing a discussion on the feeding and rationing of sheep a brief look must first be taken at how a sheep eats and processes its food.

The sheep as a grazing animal

DENTITION

In common with other mammals, sheep have two sets of teeth in the course of their lives. They begin life with a set of temporary or milk teeth which, in turn, is followed by a permanent set. Nature does not replace any of these permanent teeth when they are lost, and in consequence the loss of teeth by a sheep can have serious consequences for its owner. In both sets of teeth all the incisors are on the lower jaw and these teeth bite against a cartilagenous dental pad on the upper jaw. There are 8 incisors in both the temporary and the permanent sets of teeth.

The permanent incisors can be used as a guide as to the age of the sheep but it is by no means a precise measure. At birth a lamb may show one or two incisors up through the gum or none at all. Occasionally these milk teeth cause damage to the teats of the ewe. At about two months of age the lamb will be showing all its temporary incisors. These teeth are much smaller, narrower and more conical in form than are the adult permanent teeth. At between ten and fourteen months of age the two central temporary incisors are cast and replaced with two permanent teeth (see Fig. 2.1) which are larger, broader and more flattened in section than are the milk teeth. The temporary teeth continue to be replaced in pairs working from the inside outwards:

- At 2 years of age the sheep should have 4 permanent incisors.
- At 3 years of age there should be 6.
- At 4 years of age 8 incisors should be fully up and in wear.

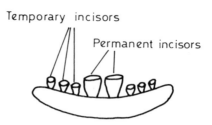

Fig. 2.1 Schematic diagram of a two-toothed sheep's lower jaw (from the front).

In addition to the 8 incisors the full permanent set of teeth of the sheep comprises 6 pairs of premolars and 6 pairs of molars on the upper and lower jaws making 32 teeth in all. In some cases a ewe will have a full mouth at the end of three years. It must again be emphasised that after the two tooth stage, teeth are less and less reliable in ageing sheep. Even sheep of the same breed, within the same flock and treated similarly will show some variation and where breed and nutrition are very variable the eruption of teeth can be equally so.

Some stages in the eruption of teeth cause problems to the animal and young sheep changing their teeth tend to have difficulties with some quite normal foods. The breaking of swedes and turnips is a case in point where the sheep have difficulty in gnawing through the 'rind' of the 'root'. Similarly heavily compacted self-feed silage can prove difficult for such sheep to handle.

Sheep tend to lose their teeth over a period of time but when these losses commence varies enormously. Various factors contribute to teeth loss, such as heredity, the type of food being eaten, the mineral status of the herbage and so on. Research into teeth loss continues, as having to cast young ewes as broken-mouthed causes serious financial losses on hill farms. Some sheep start losing teeth as early as 4 or 5 years of age while others retain a full mouth until 8 or, exceptionally, 12 years of age.

When a sheep has lost one or more of its permanent incisors it is known as *broken-mouthed*. These broken-mouthed sheep, as with young sheep changing teeth, can have trouble in eating. This is particularly so with sheep on mountain grazing and here it is normal to reject such sheep from the flock. Efficient dentures have been produced for sheep which have lost their incisor teeth but whether this will prove to be an economical solution to the problem of broken mouths remains to be seen. In older sheep broken molars can also cause trouble by lacerating the inside of the cheek.

Another condition which causes problems with eating is when the lower jaw is too short or too long relative to the upper jaw. In either case the incisors do not meet the dental pad properly (see Fig. 2.2) and, in consequence, grazing ability is impaired. The cause of malformation is congenital and both parents should be eliminated from the breeding flock, as well as the sufferer from the defect.

One further observation on teeth is that besides using the term 'broken-mouthed' to denote a class of sheep, the term *'two-toothed'* is also used to describe young sheep. Two-toothed females are also referred to as *shearling ewes* or *gimmers*, while the males are known as *shearling rams*.

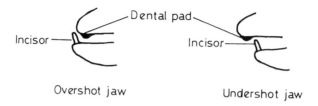

Fig. 2.2 Schematic diagram of two common jaw malformations seen in sheep — *overshot jaw* and *undershot jaw*.

THE MODE OF GRAZING OF THE SHEEP

The sheep has what is known as a split upper lip, i.e. the lip is divided down the median line, as opposed to a cattle beast which has an entire upper lip. The result is that whilst a cow rips herbage off in mouthfuls, the sheep is able to bite off individual blades of grass and to bite them off quite close to the ground. In other words, a sheep is a much closer grazer than is a bovine and it can also be much more selective in its grazing.

Two things follow from this. Firstly, when sheep and cattle are in competition for newly grown grass, the sheep will always win. It is for this reason that a flock of breeding ewes and a herd of dairy cattle are not good companions in the springtime. This gives rise to one of hill farming's problems. In some areas coastal dairy farmers take hill ewe hoggets — i.e. young sheep past the lamb stage — for wintering, but as they need to reserve the early spring grass for the dairy herd they will want to get rid of the hoggets much earlier in the spring than is opportune for the hill farmer. In other words, the hoggets must go back to the hills to compete with the ewes and lambs for grass that is, as yet, in scarce supply.

A second consequence of the sheep's mode of grazing is that in the majority of cases pastures tend to degenerate when carrying no stock other than sheep. This problem can be mitigated on low ground farms by mechanical means but these are not often practical on rough hill land; therefore the operation most commonly used on the hills to control rough growth is burning. The problem arises from the fact that especially on hill land the vegetation in summer outgrows the ability of the sheep stock to keep it under control and the coarsest of the herbage is rejected by the sheep even in winter. The following spring when regrowth starts the sheep select the growing blades of grass which are nipped off leaving the coarse material. This deterioration becomes progressive and, year by year, the pasture gets rougher. Depending on soil type and locality, such shrubs as brambles, gorse and thorn become established and unless other grazing animals are introduced or the rough herbage fired, the land reverts to wilderness. There are a few soil types with an indigenous herbage which can readily sustain grazing sheep without deterioration, a typical example being the chalk Downs of southern England.

The basic nutrition of sheep

Before feeding is discussed a brief outline of digestion in the sheep and its requirements for nutrients is appropriate. Figure 2.3 shows a schematic diagram of the digestive tract (gut) of the ruminant.

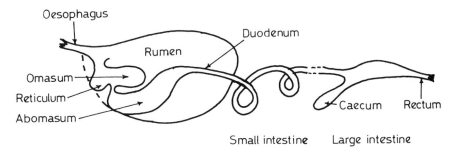

Fig. 2.3 Ruminant digestive tract.

DIGESTION

The sheep, being a ruminant, has a stomach divided into four sections. The *rumen*, in which the rumen digestion takes place, is the first and by far the largest section and is closely associated with the *reticulum*. These are followed by the *omasum* and the *abomasum*. The abomasum corresponds to the stomach in the monogastric animal and is known as the *true stomach*. When a sheep takes in a mouthful of food during a feeding session it chews it quickly and then swallows it. The ingested material passes down the *oesophagus* or gullet and into the rumen where it is attacked and broken down by the bacteria and protozoa which inhabit the rumen. These organisms produce enzymes which break down cellulose and hemicellulose. They also break down proteins in the food and incorporate the resultant breakdown products into their own bodies.

When the sheep has ended a period of feeding, it normally settles down to chew the cud, or ruminate. In this process the larger particles of food in the rumen are gathered together into balls known as *boluses* which are regurgitated. The regurgitated material is thoroughly masticated and reswallowed, then further boluses are brought up and the action repeated. At the same time the food in the rumen undergoes a churning which, in the healthy animal, is fairly continuous. The outcome of the chewing and churning is that the finer material finally passes out of the reticulo-rumen in liquid suspension into the omasum where it loses water before passing on to the abomasum for normal digestion. On leaving the abomasum the food passes to the small intestine and then to the large, undergoing processes similar to those which take place in a monogastric animal.

Rumen digestion has a number of features which need to be borne in mind when setting up a feeding regime in order to obtain a really efficient system. Volatile fatty acids, namely acetic, propionic and butyric acids are major end products of rumen fermentation.These acids represent the end products of bacterial activity and are absorbed across the rumen wall and taken into the liver along with the bulk of nutrients absorbed from the gut. Acid production is therefore proportional to food intake and to the ease with which it is broken down. They make a significant contribution to the total energy needs of the sheep.

A fraction of the dietary protein is broken down or degraded by rumen micro-organisms and this is commonly referred to as *rumen-degradable protein* or *R.D.P.* This breakdown is crucial to the micro-organisms, since it provides simple compounds of nitrogen, chiefly ammonia, which they use to synthesise their own body proteins. This bacterial protein is later digested in the abomasum and small intestine to form amino acids. Any protein that escapes breakdown in the rumen is termed *undegradable protein* or *U.D.P.* (see Fig. 2.4). Part of this will be digested alongside bacterial protein in the abomasum and small intestine and so swell the supply of amino acids made available to the animal for the synthesis of protein in new growth, in milk or in wool.

It follows that a certain minimum level of R.D.P. is necessary if the population of rumen micro-organisms is to be sustained and function efficiently. Any shortage may be made good by feeding *non-protein nitrogen* (N.P.N.) compounds as long as they can provide ammonia to the micro-organism. The most common compound used in this situation is urea and this is cheaper than using expensive protein concentrates. Care, however, is necessary to control intake since an excess could produce so much ammonia that it might prove toxic.

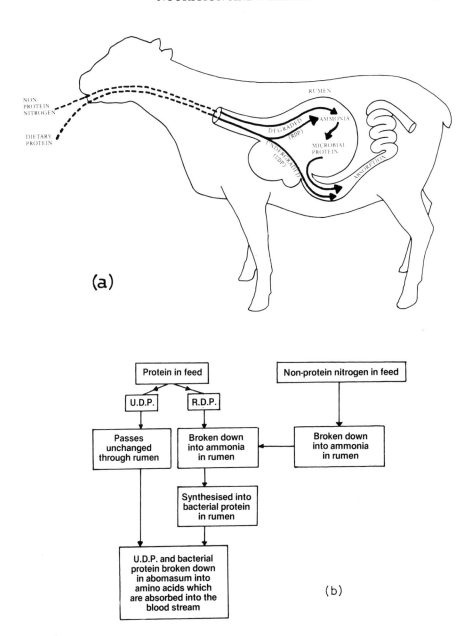

(a)

(b)

Fig. 2.4 Simplified diagrams of nitrogen metabolism in the rumen and abomasum. (*Fig 2.4(a) with permission of the Meat and Livestock Commission.*)

It is important to appreciate that although two rations may provide equal protein intakes different relative contributions of R.D.P. and

U.D.P. can alter the total supply of amino acids made available for absorption from the gut.

Rumen digestion, being a fermentation process, gives rise to heat — which can be particularly valuable in rigorous climatic conditions — and large quantities of gas. This gas is normally lost by belching, but if for any reason the gas is unable to escape the animal swells up and a condition known as *bloat* results. If not released the pressure against the diaphragm is such as to stifle the heart and cause death.

One final point should be noted on ruminant digestion. At birth a ruminant does not have this system in operation; it cannot digest cellulose and has no heat of fermentation. The young lamb behaves as a monogastric animal, as do other young ruminants, and a fold in the digestive tract known as the *oesophageal groove* by which the reticulo-rumen is by-passed ensures that the milk and other foods fed to the young animal pass straight to the true stomach and are not subjected to rumen fermentation.

It takes 14 to 21 days before the lamb eats solid food in any quantity and although some very early weaning dates have been reported, 5 to 6 weeks would seem to be a suitable minimum suckling period.

When a lamb can be satisfactorily weaned really depends on the amount and quality of the food being eaten. The food must be acceptable and also highly digestible in order that the lamb can eat a lot of it. The quantity of food an animal can eat depends on its rate of passage through the gut and this depends mainly on the digestibility of the food. This fact is relevant in the feeding of all classes of stock; animals will not consume more bulk in order to compensate for the falling digestibility and nutritional value of the food. Hoggets on autumn grass often go back in condition quite quickly because supplementary feeding has not kept pace with a deterioration in the herbage.

Having looked at how the animal deals with the foods offered to it, we must now look at the essential constituents of those foods.

Nutrient requirements

The food an animal ingests must provide it with the following necessities: energy, protein, minerals and vitamins. In addition, an animal requires water.

WATER

The provision of water for sheep often tends to be overlooked, because dry sheep on *good* pasture normally get sufficient moisture from the herbage they eat. The amount of water drunk depends largely on the dry matter intake (i.e. food minus its water content) of the animal. For most farm animals the ratio is of the order of 3 to 4 parts by weight of water to every one of dry matter eaten.

There are other factors which affect water intake. More water is required in summer than in winter. Young animals require proportionally more water than older animals, and a lactating ewe has a significantly higher water requirement than a dry one. Four litres per day should be sufficient for a medium-sized dry ewe, but substantially more is needed for one in lactation.

The best approach to the water situation is to see that it is always available and this is especially important with lactating ewes as anything that checks the flow of milk will adversely affect the total lactation production. The water offered should always be perfectly clean as sheep tend to be much more fastidious in their drinking than do cattle.

While the need to provide water for the lactating ewe kept indoors is obvious, it may not be so apparent with ewes running on pasture, but such ewes could well suffer during a dry spring or during a black frost (a black frost being one where there is no snow or rime) as in such conditions sheep can suffer considerable deprivation. When snow lies on the ground sheep can obtain the water they need by licking snow.

ENERGY

As with all animals, energy must be supplied in sufficient quantity to provide for the animal to walk about, collect food, digest it and, in fact, maintain all its bodily functions. The energy needed to do this is called the *maintenance energy requirement*. In addition, a surplus must be provided for growth and the production of milk and wool. The major portion of the energy required by sheep is derived from pasture grass, catch crops such as rape and conserved herbage such as hay. In intensive meat lamb production systems, both ewes and lambs require supplementary feeding with concentrates to secure the necessary energy intake.

The carbohydrate-rich concentrates normally used to increase the energy intake are cereal grains such as oats, barley, maize and, in areas where the 'root' is grown, dried sugar beet pulp. Grain can be fed whole or rolled, and in the case of maize the process of steaming and flaking produces an exceptionally palatable food which is especially useful for the feeding of young lambs.

In Britain there is some prejudice against feeding barley to sheep especially amongst the older generation of farmers. This is quite unjustified when one considers the number of intensively fed lambs that have been fed on barley during the past two decades. The prejudice arose, no doubt, in the days when barley was regarded as essentially a cash crop so that any barley fed to livestock would be unsaleable. The cause of its rejection by the market in those days was that, in most cases, it had become overheated or mouldy. Such grain, of whatever species of cereal, should never be fed to livestock.

PROTEIN

Protein must be provided to repair the wear and tear of body tissues, sustain growth in the case of young animals, maintain milk production in the lactating ewe and grow wool.

The chemical feature which distinguishes protein from other plant and animal tissue is that besides carbon, hydrogen and oxygen, it always contains nitrogen. The proportion of nitrogen present in protein is about 16%, and consequently the normal method of finding the amount of crude protein in a food is to determine the amount of nitrogen present and multiply by 6.25. Besides nitrogen, carbon, hydrogen and oxygen, sulphur and phosphorus are found in some proteins. The presence of sulphur in wool is an example.

The protein requirements of animals are quite variable; while a non-productive adult animal has a low protein requirement, that of a lactating ewe or growing lamb is quite high. The animal is able to make better use of dietary protein when the intake of energy is adequate.

The same herbage and succulents that provide energy normally provide adequate protein. Indeed, in flocks lambing in the spring to synchronise with the flush of grass, protein supply is rarely a limiting factor, and both ewes and lambs get ample. Legumes tend to provide more protein than other herbs, and clover or lucerne (alfalfa) hay can make a very useful contribution to the diets of housed sheep. Clover, especially white clover (*Trifolium repens*), has another feature which is of importance to the grazing animal, especially fattening lambs: the digestibility of the plant falls more slowly than does the digestibility of the companion grasses.

Concentrated protein foods for sheep in Europe are normally

based on peas, beans and oil seed residues. Of these, soya bean meal is a particularly useful feed. Fishmeal is an excellent source of protein which has the additional advantage of being high in essential minerals and also in having low rumen degradability (see earlier section on Digestion).

NON-PROTEIN NITROGEN

Non-protein nitrogen can be used by the micro-organisms that live in the rumen to synthesise their own protein which, in turn, is digested later in the abomasum. The normal way of providing non-protein nitrogen is to feed urea. The usual practice is to use a proprietary preparation, which is commonly a block of material based on maize, or a similar meal, and molasses into which urea has been incorporated. The sheep obtain the urea by licking the block. A liquid preparation of molasses and urea can be used for sheep fed indoors. The liquid is placed in a tank which is covered except for a small opening in the cover which is closed by a floating ball made of plastic. The sheep lick the ball which revolves under the licking and is thereby replenished. Under both systems care must be taken as some sheep tend to over-indulge, which can lead to ammonia toxicity.

The use of urea and similar nitrogenous compounds to supplement low-protein diets has proved satisfactory in practice but a little care is needed to achieve good results. It is not the best method of providing nitrogen for deep-milking ewes at the height of their lactation. Here it is better for the protein supply to be based on such low rumen-degradable foods as fish meal or soya beans and these foods should be used for indoor feeding or to outdoor ewes when the grass is not yet fully productive.

MINERALS

Most of the minerals consumed by sheep are constituent elements of the herbage they eat. The type of plants present in any situation depends to a large extent on the mineral status of the soil. In the days when horses were of more economic importance on farms than they are today, there were areas famed for 'good-boned' horses; in other words, the soil of these places contained ample available calcium and phosphorus. In addition to the useful mineral elements in plants, plants can also contain minerals which, when in excess, can act as poisons; typical of such elements are molybdenum and selenium.

There are a number of minerals which are required by animals for their natural functioning such as building bones and teeth and the formation of body fluids including saliva. The incidence of deficiency diseases varies from country to country and area to area depending on soil formation and other factors. In Britain, the main deficiencies which arise are of calcium and phosphorus and these occur in their most serious forms in the highland areas of high rainfall and poor soil parent material.

The following notes illustrate the kinds of deficiency and imbalance that can be met with in the field and against which preventive measures may need to be taken.

Calcium

Calcium is a major constituent of the skeleton and teeth. Sufficient calcium must be provided (see Table 2.1) or a deficiency of calcium in the diet of a young animal will result in a malformation of the bones known as *rickets*. The condition can also be brought about by phosphorus deficiency or imbalance, or by lack of Vitamin D.

A temporary inability to mobilise calcium by the lactating female results in *hypocalcaemia* (which is discussed in Chapter 3). A prolonged calcium deficiency in older animals produces *osteomalacia*, a condition where the bones are readily broken, but lack of phosphorus is also involved as an important cause of this condition. The sheep's normal source of calcium is from green herbage, especially legumes. Supplementary calcium is usually provided in the form of steamed bone flour, ground limestone or fish meal.

Table 2.1 Dietary requirements (g/day) for calcium and phosphorus for growing sheep.

Liveweight (kg)	Daily liveweight gain (g)	Ca	P
30	200	3.7	2.1
35	220	4.1	2.3
40	250	4.6	2.6
45	250	4.8	2.7
50	220	4.4	2.6
55	200	4.3	2.6

Reference: ARC (1980)

Phosphorus

A deficiency of this element can also cause rickets or osteomalacia. Phosphorus is found in the skeleton in combination with calcium, but it is also present in other parts of the body in quantity. Phosphorus deficiency, where severe, often gives rise to *pica* or depraved appetite. As with other minerals, a sheep gets its phosphorus from herbage but this is not a source of abundant supply. Cereal grains are fairly high in phosphorus content but in many cases it is of low availability. The normal source of phosphorus for use as an additive in concentrate foods is steamed bone flour, fish meal, and blood and bone meal.

As previously mentioned calcium/phosphorus imbalance in a diet can cause trouble. The appropriate calcium/phosphorus ratio lies between 1 : 1 and 1 : 2 (see Table 2.1).

Sodium and chlorine

These are taken together as the deficiency of either is usually corrected by the addition of sodium chloride (common salt) to the ration. Sodium, together with potassium and chlorine, are concerned with the control of osmotic pressure in the body fluids. Chlorine is required for the synthesis of hydrochloric acid used in the digestive system. Salt deficiency causes loss of appetite but is an uncommon condition amongst sheep. The provision of rock salt licks was probably the first mineral supplementation made by farmers and this is still an appropriate way of providing salt. Most animals are attracted to common salt and it forms part of most mineral mixtures used in supplementary feeding. These substances just discussed — calcium, phosphorus, sodium and chlorine — form the basis of most common mineral mixtures, the compounds being steamed bone flour, precipitated chalk and common salt. Such mixtures are commonly incorporated in the concentrate ration at the rate of 3% by weight.

Besides powdered mixes, minerals can also be bought as proprietary bricks known as salt licks. These often contain other essential elements, e.g. cobalt and iodine, but the use of expensive mineral licks should not be embarked upon without careful consideration, there being no point in providing elements which are not in short supply.

Magnesium

This is a mineral associated with calcium and phosphorus in the skeleton and is also present in the body fluids. A deficiency of magnesium in the blood is known as *hypomagnesaemia* or *magnesium tetany*, and is discussed in Chapter 3. The condition is prevented by feeding magnesium oxide or by dressing pastures with magnesium limestone.

Sulphur

This is an important constituent of the animal body and occurs in various proteins. It also occurs in some vitamins. Urea, which is commonly fed to sheep as a non-protein nitrogen source, requires sulphur for its proper utilisation. The wool of the sheep contains substantial amounts of sulphur and there are conditions where sulphur deficiency can occur. In western Europe and other industrialised countries pollution of the atmosphere, and therefore rain, with sulphur dioxide ensures that the danger of sulphur shortage is minimal.

The minerals we have discussed are all required in relatively large amounts. There is, however, a further group each member of which is necessary for good health but needed only in very small quantities. These are known as minor or *trace elements*.

Copper

This trace element is necessary for the formation of haemoglobin in blood. Iron is also necessary for the same purpose but iron is so ubiquitous and present in such quantities that iron deficiencies rarely arise in sheep or lambs. Under most conditions copper is provided in sufficient amounts from normal foods.

In some areas such as on peat land, the herbage is low in copper. In Britain the principal disease associated with low copper is *neonatal ataxia* or *sway-back*. In this condition, the lambs are born with the brain not fully developed and the extent of this failure to develop can be such that the lamb cannot stand up, while in other cases it may show as not more than a poorly co-ordinated gait. These sway-backed lambs are all born to ewes with low blood copper, but not all ewes with low blood copper give birth to lambs with swayback. A similar condition is found in Australia where it is known as *enzootic ataxia*. Indeed, problems of copper deficiency are worldwide and particular black spot areas occur in South Africa, New Zealand and South America. There is a marked genetic component in the susceptibility to sway-back, some breeds being more prone to it than others. In some areas where the disease occurs, the soil and herbage is copper deficient; in others there is no obvious copper shortage in the soil but blood coppers in the ewes are low.

Peatland areas which have been reclaimed and heavily limed are

particularly inclined to produce the problem when grazed by pregnant ewes. It has been noticed amongst hill sheep in Britain that
sway-back is also more prevalent in years when there has been little
snow and when snow has not lain long on the hills. This phenomenon
has not been fully explained but appears to be due mainly to the
absence of the hand-feeding which normally accompanies heavy and
prolonged snowfall, the copper present in the hay or concentrates
being enough to boost the blood copper of the ewes above danger
level.

There is no cure for sway-back but the condition can be prevented
by administering copper sulphate solution to the ewes in later
pregnancy. Mineral licks containing copper can prove very effective
but quite often they are rejected by the sheep. Any flockmaster
whose sheep are troubled with sway-back should seek veterinary
advice, especially with regard to dosing with copper sulphate solution, as copper is a serious cumulative poison in sheep. Copper
injections under veterinary advice are now becoming the standard
preventive treatment. Copper poisoning in sheep on pasture has
been reported from Australia but in Britain it has occurred mainly
in housed lambs. The copper builds up in the liver as in a reservoir
and ultimately overflows into the bloodstream. When this happens
deaths amongst lambs are both sudden and numerous.

Great care should be taken on farms carrying both sheep and
swine to ensure that no pig fattening meals are ever fed to sheep.
Many modern pig fattening mixtures are fortified with copper salts
and these could rapidly prove fatal to lambs. By the same token
manure from pigs that have received copper supplements should not
be used to top dress pastures which are to carry sheep as it is possible
that the sheep could be endangered. Sheep farms agisting stock
should likewise check against pig farm residues.

A genetic factor is involved in the incidence of copper poisoning.
Some breeds are much more prone to the complaint than others,
the North Ronaldsay seaweed-eating sheep being particularly vulnerable.

Another element which enters into the copper problem is molybdenum. Molybdenum occurs in relatively heavy concentrations in
some calcareous soils. In England the Lower Lias geological formation in Somerset is the classical home of the *'teart'* pastures. Similar
conditions exist on volcanic soils in New Zealand. The high molybdenum levels in the soils are reflected in the herbage, especially in
legumes. The excess molybdenum seems to interfere with the uptake

and retention of copper resulting in an induced copper deficiency. The condition, known as 'teart', shows in sheep as scouring and general unthriftiness. It can be corrected by the provision of copper sulphate. Molybdenum is itself an essential trace element but its deficiency does not appear to cause problems in sheep.

Cobalt
This element is of great importance to ruminants. It is a constituent of Vitamin B$_{12}$ which is synthesised in the rumen by micro-organisms. A deficiency of cobalt lends to a wasting disease known as *pine* (the animals virtually pine away), *vanquish*, *bush sickness*, and by many other names. The symptoms can be confused with Johne's disease, worm infestation and other wasting conditions. The condition is a problem of world-wide incidence but, fortunately, it is now readily controllable. Cobalt sulphate can be used to treat the land. The salt is diluted with a fertiliser such as superphosphate of lime and applied at the rate of 2 kg cobalt sulphate per hectare.

Cobalt sulphate can be given by mouth in solution but now the most popular method of direct dosage is to administer cobalt 'bullets'. 'Bullets' of a cobalt compound mixed with clay or a similar base can be lodged in the reticulum where they release small amounts of cobalt over a long period of time, obviating the necessity for frequent treatments. Cobalt can also be supplied as a mineral lick. The areas deficient in cobalt can vary widely in size; some areas are huge while others may be small fields. In Britain the disease occurs in the Scottish Borders, the Isle of Tiree and many other places.

Selenium
This is an essential trace element whose deficiency is associated with *muscular dystrophy*, *infertility* and *poor wool growth*. It has a relationship with Vitamin E as either can be used under certain conditions to prevent muscular dystrophy in sheep and cattle, but the relationship is not fully understood. In Britain selenium deficiency has not constituted a serious problem to date, but in some countries such as New Zealand and Canada the shortage can be serious.

Selenium can be administered as a constituent of a mineral mix, but veterinary advice is necessary as selenium is a toxic substance. Indeed, in some areas of the world this poisoning effect is serious. Selenium poisoning is common in some areas of the United States where it is known as *alkali disease*, or *blind staggers*. The selenium acts by displacing sulphur in the cystine bridge in the proteins of hair and horn causing the sloughing off of hooves, the changing of hair colour, and similar degenerations in the animal. Selenium poisoning seems to be associated with the eating of certain weeds which have a high uptake of the element. An elimination of such plants and the feeding of high protein concentrates offers the best line of defence against the condition.

These, then, are the minerals of which sheep farmers, at present, need to take note. Changes in farming or industrial practices could, of course, give rise to other problems and farmers should never adopt the view that 'we know it all'. In recent years contamination of herbage by smoke containing fluorine has, for example, caused problems of tooth decay in sheep.

VITAMINS

As the sheep is essentially an outdoor animal and a ruminant, it rarely suffers from vitamin deficiency (other than B_{12}). There are, however, a number of situations where problems can arise. The commonest situation giving rise to these is the long-term housing of sheep and here the vitamins most likely to be lacking are those of the fat-soluble group – A, D and E.

Vitamin A

While a shortage of this vitamin is not commonly met with in sheep, it is quite possible for a deficiency to occur in housed sheep kept throughout the winter on a diet of poorish hay plus concentrates. The condition shows itself in night blindness and in lambs being born weakly or dead. Up to fifty years ago the condition was also quite common in housed cattle. Good quality green hay or silage could prevent the condition as would green soilage; failing these preventatives, the stabilised synthetic vitamin should be incorporated in the concentrate ration. On pasture, the disorder does not occur as carotene, the precursor of the vitamin, is present in all green herbage.

Vitamin D

The two important vitamins of the D group are Vitamins D_2 and D_3 – *ergocalciferol* and *cholecalciferol*. A lack of these vitamins produces *rickets* and *osteomalacia*. The vitamins are formed by the action of ultra-violet light on ergosterol and dehydrocholesterol and both have similar effects on the animal. Animals get the major portion of their Vitamin D requirements via the action of the sunlight on the dehydrocholesterol contained within the skin of the animal, particularly the more exposed parts such as the face, ears and legs. It is for this reason that the deficiency is most commonly found in housed animals.

Hill sheep on north-facing hirsels can also be short of the vitamins in years of perpetually low cloud and little sunshine. While some Vitamin D is contained in well-harvested, sun-dried green hay, it is advisable to feed a Vitamin D supplement to housed pregnant or lactating ewes. This is best done by incorporating the synthetic stabilised vitamin in the concentrate diet. Vitamin D (which the animal stores in the liver) can also be given by injection and can be used with hill ewes in a bad winter, but it should be borne in mind that over-dosing could cause Vitamin D toxicity. It follows that where injection of Vitamin D is considered veterinary advice should be sought.

Vitamin E

This is really a group of substances known as tocopherols, the most active of which is α-tocopherol. Deficiency of the vitamin is associated with muscular dystrophy and also infertility. The animals' main source of α-tocopherol is green herbage and cereal grain, wheat germ oil containing a substantial quantity. A synthetic product is available for the treatment of deficiency conditions and this is now becoming an integral part of rations for housed sheep.

Water-soluble vitamins: Vitamins B and C

Vitamins of the B group are synthesised in the rumen and therefore do not need to be supplemented in the sheep's diet. Vitamin B_{12} (cyanocobalamin) deficiency occurs in the absence of dietary cobalt as has been explained. Vitamin C is synthesised metabolically by all farm animals hence dietary supplementation is unnecessary.

PALATABILITY, ACCEPTABILITY AND OTHER PROPERTIES

The preceding paragraphs have outlined the various groups of substances which are necessary for the proper nutrition of sheep. Additional mention has been made of a number of minor elements which can cause problems either by their absence from the diet or from over-abundance. Our next requirement is to give some indication of the sources of these foods and the quantities in which they are required by the animal. But before this is done, attention must be drawn to a number of general considerations which must be taken into account when arranging rations for livestock.

The first point is that it does not matter how well a diet is compounded in respect of its nutritional status if the animals in question will not eat it in the desired quantities. This means that the farmer must pay attention to *palatability* and *acceptability*. A food may be rejected by the animal because it is mouldy, has been contaminated by some chemical or is in a physically unacceptable condition. For example, lambs may reject concentrate pellets because they are so hard as to make the physical act of eating difficult. On the other hand, a concentrate meal may be so finely ground that it becomes pasty in the mouth and becomes unattractive to the animal.

Sheep, like other animals, may reject food for no better reason than that it is unfamiliar. This response often makes feeding of supplementary concentrate food to extensively grazed sheep difficult. Where the customary supplement has been hay but conditions arise where concentrates have to be substituted, the latter food is often rejected. Similarly with fattening hoggets, those whose mothers were trough-fed at lambing adapt to a concentrate diet much more readily than do those whose mothers were not.

Returning to the actual feed, it must contain sufficient nutrients for the function it has to perform, i.e. promote growth, provide milk, grow wool and so on. To take the example of an ewe in late pregnancy carrying twins, feeding a sole diet of low-quality hay is courting disaster. In addition to its nutritive contribution the food must have no ill effects on the animal. Rape and kale provide good examples of such feeds; the feeding of these crops must be controlled or poisoning may result. Finally, the ration must be *economical*. There is a natural tendency for anyone who has found a satisfactory feeding regime to continue with it from year to year and this can have its dangers when price fluctuations are extreme. A case in point is where the farmer normally buys in hay for winter feeding. In years of adverse hay harvesting weather it is often more economical to replace some of the hay with sugar beet pulp or barley. The same consideration applies to concentrates and while these should not be changed in response to small price fluctuations, large price changes should be countered by reconstructing the rations.

The supplementary feeding of sheep

In most sheep farming situations other than indoor and out-of-season lambing and rearing it is assumed that the sheep will get the major portion of their sustenance from pasture grasses, legumes and other herbage. There will, however, be times when supplementary feeding is necessary or desirable. It is important to ensure that such provision is made economically. Under these conditions the farmer with mixed enterprises is at an advantage. He can make use of arable residues in many cases. He can devote part of his arable land to the growing of supplementary crops such as 'roots' or forage. He has the machinery for soil preparation, seeding and harvesting crops.

The marginal land or hill farmer, on the other hand, is in a much different position. He usually has a relatively small acreage of tillage land and it is unwise for such a person to invest in a lot of expensive cultivation machinery. In many areas the problem of cultivation can be overcome by the use of contractors. Having given a reminder of the importance of the economics of scale, a few notes on crops may now prove helpful.

On farms other than arable or large mixed farms where cereals are normally grown, concentrate energy feeds are best bought in. The same applies to high protein concentrates such as legume seeds and oil seed residues. The question of forages is more difficult. For instance, buying in hay to hill and marginal land areas usually implies a very heavy transport charge. On the other hand the problems of growing and especially harvesting hay in hill areas such as those of northern and western Britain are great. Ill-won, mouldy hay is not the right food for sheep, especially in-lamb ewes. The resolution of this problem must be left to the individual farmer but some are beginning to solve it by making silage. Making silage for sheep is similar to making it for cattle but the importance of achieving a product of high digestibility with a high dry matter content is greater for sheep than for cattle. Great care must also be taken to ensure that the cut crop is free from extraneous matter such as soil from mole hills, and the setting of the forage harvester must be adjusted accordingly. A high setting of the cutter bar also ensures that most of the coarse, indigestible portion of the herbage is left behind. Sheep can be self-fed from the silage clamp but the settled height of the silage should not exceed 1.25 metres.

CROPS FOR FEEDING *IN SITU*

In most sheep keeping areas some crops are grown specifically for eating off by sheep. The majority of these are either legumes or brassicas. While it is not the purpose of this book to go into arable crops in any depth, there are a number of points to which sheep farmers inexperienced in arable farming should be alerted.

The environment is even more important to a plant than to an animal as, within reason, an animal can get up and go to another and better one. It can usually move out of a draught, move to a water supply and, in the case of a range animal, move from a poor grazing area to a better one.

The plant has no such option. It has to stay put. Therefore the farmer has to supply as good an environment for the plant as is possible. He must ensure that the soil contains adequate nutrients, water, has a suitable pH and does not become water-logged over long periods, and so on. He must therefore arrange to have a soil analysis in order to see what quantities of fertiliser and lime he requires. It is not enough to know that a soil is acid; the farmer also needs to know the lime requirement necessary to achieve a satisfactory pH. It may be that the requirements for lime and fertiliser are such as to make the enterprise economically doubtful.

Furthermore, plants − like animals − can suffer from minor element deficiencies; e.g. boron deficiency causes a serious disease in cruciferous crops, especially swedes. There is the additional risk of damage from transmissible diseases and insect pests. Examples from brassica crops such as kales and turnips are the fungus disease *finger-and-toe disease* (caused by *Plasmodiophora brassicae*) and *turnip flea beetles* (*Phyllotreta* spp.) which attack emerging seedlings. Finger-and-toe disease can cause serious loss in soils heavily infected with the organism while the whole crop can be lost due to flea beetle attack when the seed has not been dressed with an appropriate seed dressing.

Legumes

These plants have two noteworthy features:

- Most crop plants of this family cannot thrive in acid soils.
- They are able to fix atmospheric nitrogen due to the presence of specific bacteria which live symbiotically in their root nodules. Where seed of these legumes is to be sown on soil that has not previously carried the crop it is necessary to inoculate the seed with the appropriate strain of bacteria.

A particularly useful legume on deep soils in the drier areas of the country is lucerne (alfalfa). One of the major characteristics of lucerne is its ability to withstand drought. It does this by means of a long, deep tap root. In the initial stages of growth this causes problems as the young plant devotes most of its energy to putting down this tap root, and as a result surface growth is slow and the plant competes badly against weeds in the early stages of life. When, however, it is established it is an excellent competitor. It follows that when a lucerne ley is to be established the land needs to be exceptionally clean and also well-fertilised.

Turning to legumes such as red clover these, too, have their problems. Land can become clover-sick due to a fungus, *Sclerotinia trifolium* which kills the plant. Clover sickness can also be brought about by heavy eel-worm infestation which destroys the root system. In these circumstances alternative crops must be sought.

The above is enough to show that there can be snags to growing crops for the supplementary feeding of sheep but it is not intended to suggest that only the most gifted can grow them. The prospective grower must be aware of possible difficulties in order that the proper precautions against failure are taken. A sheepman in a new farming situation and thinking of introducing a new crop should take local advice from both farmers and advisory agencies such as A.D.A.S. or College Extension services. The fieldsmen of these organisations have detailed knowledge of their districts and are in a position to help the farmer avoid dangers which, although real, may not be obvious.

Having looked at basic nutrition and the sorts of food fed to sheep we must now turn to rationing.

Sheep rationing

METABOLISABLE ENERGY

Having discussed the qualitative side of nutrition and the sources of various nutritional elements it is necessary to put some figures to feeding. The practical farmer will ask such questions as: 'How much of so and so do I need to feed to achieve a given result in a stated time?'. He may, for instance, want to know how much food a certain type of ewe, thought to be carrying twin lambs, requires in order that she arrives at parturition in optimum body condition. On the other hand his question may be: 'I am feeding so much oats per head per day but need to change to barley. What is the replacement value of barley relative to oats?'.

Time and space forbid the story of how research workers have tackled the problems of putting figures to ideas and working out a quantitative method of stating the nutritional requirements of various animals. Students should nevertheless consult an authoritative book such as *Animal Nutrition* or *Improved Feeding of Cattle and Sheep* (see Book List) for a full explanation.

The sheep needs a daily supply of suitable food to provide enough energy to maintain its body functions, produce milk and wool, grow and put on flesh, and also sustain the process of pregnancy. Foods containing carbohydrates such as sugars, starches and cellulose, provide the main source of energy, but energy is also obtained from fats and proteins. Carbohydrates, however, are the main source of energy to ruminants.

Energy is measured in *joules*; the joule is now the international unit of energy and replaces the calorie (1 calorie = 4.184 joules). A convenient unit in animal nutrition is the *megajoule* (1 MJ = 1 000 000 J).

The modern approach is to measure the energy content of the feed at different stages of use (and non-use) by the animal. The starting point is the *Gross Energy* i.e. the total energy in megajoules which would be released as heat if the food were completely burned. The gross energy present in any food can be determined quite readily by calorimetry but this alone is not very helpful as some of the energy-providing content is not digested but is voided in the faeces. Only part of what remains – the *Digestible Energy* – can be utilised, however. Some is lost in the urine (as *Urinary Energy*) and some more as methane gas (*Methane Energy*), the latter being a highly significant loss from the sheep's gut. The *Metabolisable Energy* is what is left and is the energy extracted from the feed which is actually useful to the animal (see Fig. 2.5). Metabolisable energy is used for:

Maintenance (including general activity)
Production (such as lactation)
Growth
Pregnancy

It is important to note that metabolisable energy (M.E.) can be divided into (a) the *Heat Increment* (i.e. the heat used up during the conversion of the feed into a useful form by the body) and (b) the *Net Energy* (i.e. the net residual energy available for use in maintenance, production, growth etc). The maintenance energy is eventually lost to the atmosphere as heat, along with the heat increment, but it has in the process kept the animal at a sufficiently warm

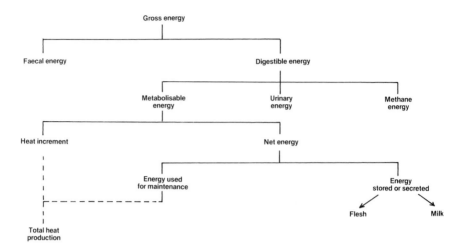

Fig. 2.5 Partitioning of feed energy within the ruminant (*adapted from ARC Technical Bulletin 33*).

temperature. Sheep are warm-blooded animals and their body processes will not function if their body temperature falls below a certain critical value.

Evidently, the metabolisable (i.e. utilisable) energy is a crucial figure for working out rations, but as various feeding stuffs have very different moisture contents, calculations have to be made on a dry matter basis. The figures given in tables of food composition are therefore in megajoules per kilogram of dry matter, MJ/kg D.M. (See Table 2.3).

Full details of how rations are calculated on the above basis can be found in the Ministry of Agriculture's Bulletin 33, *Energy Allowances and Feeding Systems for Ruminants.*

Rations can be compounded on the basis of this system without a detailed knowledge of the calculations. Tables and graphs such as those published by the Meat and Livestock Commission can be used and from these energy and protein requirements of animals in different stages of production can be extracted.

It is also quite easy for the person rationing ruminants to make changes in various constituents by reference to energy and protein composition tables for various foods. In addition to the tables of energy and protein values of food stuffs the flockmaster needs to know the weight of his animals as the amount of food an animal will eat is proportional to its live weight. The average ruminant will consume about 2.5–3.0% of its live weight in dry matter per day. The heavily lactating ewe appears to share with the dairy cow the ability to increase this capacity from about a week after lambing until her milk yield passes its peak. However, there will be a period before parturition when her dry matter capacity will fall off. This is particularly noticeable when a ewe carries two or more foetuses. In order to keep up the nutrient intake as far as possible the quality of food must be improved. This means feeding more concentrates and it is also at this time that the highest quality hay should be fed.

Table 2.2 is taken from the Meat and Livestock Commission's published pre-tupping weights of some of Britain's commoner breeds and crosses.

Table 2.2 Mature ewe body weights in kg.

(a) Hill breeds			
South Country Cheviot	48	Swaledale	45
Scottish Blackface	50	Welsh Mountain	32
Kerry Hill	47		
(b) Long wools			
Border Leicester	86	Devon and Cornwall longwool	82
(c) Down breeds			
Dorset Down	75	Oxford	98
Southdown	57	Suffolk	82
(d) Short wools			
Clun Forest	63	North Country Cheviot	82
Devon Closewool	57	Dorset Horn	73
(e) Crossbred fat lamb mothers			
Greyface	70	Mule	69
Masham	72	Welsh Halfbred	59
Scottish Halfbred	75	Suffolk × Scottish Halfbred	82

This table can be used as a guideline but, when feeding ewes, body score, stage of pregnancy and similar conditions must be taken into account.

The normal rationing procedure is to calculate the quantities of food required to provide the necessary energy and then to adjust the ration by the addition of a protein-rich source to get the appropriate energy:protein balance.

The farmer usually does not calculate minerals and vitamins when he decides on his ration. What he does is to bear in mind the mineral status of the various component parts of the total ration and make adjustments. These adjustments are made with reference to the type of production in mind. The animals with the greatest mineral needs are lactating ewes and rapidly growing lambs. Lactating ewes feeding twins or triplets will need supplementary mineral feeding if the milk yield is not to fall or too big a demand is made on the ewe's own body tissues. This is particularly true of housed ewes. The normal method of achieving this is to feed a 3% mineral supplement in the concentrate ration which can be reduced to 2% if the concentrate ration has a reasonable fish meal component.

The Vitamin A and D requirements are best met by incorporating a proprietary vitamin mix into the concentrate ration at the rate recommended by the manufacturers.

COMPOSITION OF FEEDING STUFFS

Table 2.3 gives the composition of some of the commoner feeding stuffs, together with the percentage degradability of the protein. The energy figures have been taken from the MAFF Technical Bulletin No. 33.

In the table below there are a few points to which particular attention should be paid. The first is the relatively high M.E. values of the dry matter in roots and tubers. The second is the rapidity with which the digestible crude protein fraction falls off as the grass crop matures: the available protein in seeded hay is less than half that of leafy material. Finally, fish meal can make an outstanding contribution to the calcium and phosphorus content of a ration.

It will be noted that the values for protein are given in grams of digestible crude protein (D.C.P.). There is a possibility that the scientific world may move to quantify the R.D.P. and N.D.P. contents in future but, in view of the need for further research and development work, present estimates must be treated as provisional. However, the author feels that the concept of D.C.P. will be around for a long time yet and that the ability to quantify what happens to nitrogen in the digestive tract of a sheep is not of paramount importance to sheep men provided the utilisation of the two types of protein is kept in mind. This is the reason for staying with the D.C.P. system, but before going on to give examples of rationing a few more words will be said about the rumen degradability of protein. Although it is not so important as in the nutrition of the high yielding

Table 2.3 Food composition values per kg of dry matter.

	Dry matter (g)	Metabolisable energy (MJ)	Digestible crude protein (g)	Ca (g)	P (g)	Mg (g)	Degradability of protein (%)
Food grains							
Oats	860	11.5	84	0.8	3.4	1.0	85–95
Barley	860	13.7	82	0.7	4.0	1.0	85–95
Maize	860	14.7	78	0.6	5.8	2.6	60–70
High protein supplements							
Soya bean meal (extracted)	900	12.3	453	3.6	7.5	3.0	45–65
Ground nut meal (decorticated)	900	11.7	491	2.2	7.1	0.4	65–85
Fish meal	900	11.1	631	68	38	2.1	25–45
Roots							
Mangels	125	12.4	58	1.9	1.9	1.9	85–95
Swedes	120	12.8	91	6.4	2.2	1.0	85–95
Sugar beet pulp (dried)	900	12.7	59	5.9	0.6	1.0	80–90
Potatoes	210	12.5	47	2.1	4.4	3.0	85–95
Dried grass							
Very leafy	900	10.8	113				
Leafy	900	10.6	136	6.0	3.3	1.6	75–90
Early flower	900	9.7	96				
Hay							
Clover	850	9.6	128	9.6	2.8	4.5	35–65
Ryegrass leafy	850	10.1	90				
Ryegrass early flower	850	9.0	58	3.0	2.1	0.8	70–80
Ryegrass seed set	850	8.4	39				
Barley straw	860	7.3	9	3.4	0.9	1.9	80
Grass silage							Untreated
Leafy (high digestibility)	200	10.2	116				80–90
Early flower (medium digestibility)	200	8.8	102	5.0	3.1	1.6	Formic acid added
Full flower (low digestibility)	200	7.6	98				30–50
Molasses	750	12.8	44				100

dairy cow it can be of some moment to the lactating ewe, especially the ewe kept indoors. As Dr J. J. Robinson and co-workers of the Rowett Research Institute have shown, the ability of the ewe to mobilise her own body fat and utilise it for milk production is strongly influenced by the availability of rumen undegraded protein. In order that this aspect of the ewe's nutrition can be catered for, a ration which does not contain a reasonable amount of undegradable protein should be modified in good time. If, for instance, ewes are being fed a concentrate ration where the protein function is supplied by decorticated groundnut cake a substitution of part of this by, say, fish meal should be started some three weeks before lambing and the substitution completed, say, a week before lambing. The substitution should take place gradually as should all changes in diet of all animals. If the change is made suddenly it may result in the total rejection of the concentrate ration. This rejection not only has a nutritional effect but it also acts as a stress factor. The stress of lambing is quite large enough without adding to it. Finally, it should be emphasised that the better the quality of the energy supply the more efficiently will the protein be used.

Tables 2.4, 2.5 and 2.6 give the energy requirements for different sheep under different conditions and on the basis of these and protein requirements satisfactory rations can be calculated.

Table 2.4 summarises Technical Bulletin 33 recommendations for outdoor ewes, which also suggest a deduction of from 0.7 to 1.2 MJ/day for indoor ewes depending on body weight.

Bulletin 33 advises that an addition of from 0.3–0.7 MJ per day be added to the ration of outdoor lambs depending on their weight but, in the majority of cases, this is unlikely to be necessary.

A perusal of the last three tables will underline what every successful shepherd knows, namely that the nutritional requirements of a milking ewe in early lactation are very high. It also underlines the fact that hoggets will not put on flesh with a limited quantity of poor quality food.

Knowing the requirements of particular animals for energy it remains to find those for protein. The following Tables 2.7 and 2.8 are based on the recommendations of the East of Scotland College of Agriculture.

Given the weight of the animal and the particular production envisaged, together with the figures for energy and protein requirements, the feeder is in a position to produce a suitable ration.

The practical farmer is not normally faced with the academic

Table 2.4 M.E. allowances for ewes (MJ/day).

Live weight (kg) and single (S) or twin (T) pregnancy		Dry non-pregnant	Weeks before lambing				
			8	6	4	2	0
30	S	4.8	5.1	5.7	6.3	6.9	7.7
	T		5.1	5.9	6.8	7.9	9.2
40	S	5.8	6.1	6.7	7.4	8.2	9.1
	T		6.1	7.1	8.2	9.5	11.0
50	S	6.8	7.0	7.8	8.6	9.5	10.5
	T		7.1	8.3	9.6	11.1	12.8
60	S	7.8	8.0	8.8	9.8	10.8	11.9
	T		8.1	9.4	10.9	12.7	14.7
70	S	8.8	8.9	9.9	10.9	12.1	13.4
	T		9.2	10.6	12.3	14.2	16.5
80	S	9.8	9.9	10.9	12.1	13.4	14.8
	T		10.2	11.8	13.7	15.8	18.3

question of what is the most desirable ration that can be devised for a particular group of sheep. The case is usually that he has x tons of hay or silage to last the winter, or he has so much hay and so many acres of swedes for feeding off a flock of hoggets. In such cases what he needs to do is to find out how much fodder he can afford to feed per day and how much concentrate, if any, he needs to balance it. In practice as well as in theory a farmer should have his fodders such as hay and silage analysed if this is at all possible. If not he should work on the basis that it is probably not as good as he thinks it is! This is especially so in respect of protein content.

Example 1

To devise a ration for a group of mule ewes average weight 70 kg in their eighth week before lambing, of which the majority, hopefully, are carrying twins.

The dry matter capacity of each ewe would be about 2.5% or 2.50 × 70 = 1.75 kg D.M./day. From Table 2.4 the M.E. for maintenance and pregnancy is 9.2 MJ/day. If we take a general traditional

Table 2.5 M.E. allowances for lactating ewes (MJ/day).

Live weight (kg) and single (S) or twin (T) lambs		Stage of lactation (months)					
		Hill ewes			Lowland ewes		
	1	2	3	1	2	3
30	S	15.3	14.9	12.2	–	–	–
	T	22.3	19.2	14.6	–	–	–
40	S	16.3	15.9	13.2	–	–	–
	T	23.3	20.2	15.6	–	–	–
50	S	17.3	16.9	14.2	19.3	18.9	15.7
	T	24.3	21.2	16.6	26.6	23.2	18.0
60	S	18.3	17.9	15.2	20.3	19.9	16.7
	T	25.3	22.2	17.6	27.6	24.2	19.0
70	S	–	–	–	21.3	20.9	17.7
	T	–	–	–	28.6	25.2	20.0
80	S	–	–	–	22.3	21.9	18.7
	T	–	–	–	29.6	26.2	21.0

Reference: Technical Bulletin 33.

Table 2.6 Net Energy allowances (MJ) of growing lambs (indoors).

L.W.G. (g/day)	Live weight (kg)								
	10	15	20	25	30	35	40	45	50
50	2.4	2.9	3.4	3.9	4.4	4.9	5.4	5.9	6.4
100	3.2	3.7	4.2	4.8	5.3	5.8	6.4	6.9	7.5
150	4.0	4.6	5.2	5.7	6.3	6.9	7.5	8.1	8.7
200	4.9	5.5	6.1	6.7	7.3	8.0	8.6	9.3	9.9
250	5.8	6.4	7.1	7.7	8.4	9.1	9.8	10.5	11.2
300		7.3	8.0	8.8	9.5	10.2	11.0	11.7	12.5
350			9.0	9.8	10.6	11.4	12.2	13.0	13.8
400				10.9	11.7	12.5	13.4	14.3	15.2

Reference: Technical Bulletin 33.

Table 2.7 Allowances of Digestible Crude Protein (D.C.P.) for dry, pregnant and lactating ewes.

Live weight (kg)	Mainten-ance D.C.P. (g/day)	Late pregnancy		Lactating	
		Single	Twins	Single	Twins
40	48	66	87	201	286
50	56	88	101	209	294
60	63	99	115	216	301
70	69	110	127	222	307
80	76	120	140	229	314

Table 2.8 Digestible Crude Protein (D.C.P.) allowances for growing sheep.

Daily gain (g)	100	200	300
Live weight (kg)		D.C.P. (g/day)	
20	60	87	113
30	67	94	120
40	73	99	126
50	81	107	132
60	88	114	139

ration of 1 kg hay and 2 kg swedes a day, this gives:

	D.M.I. (kg)	M.E. (MJ)
1.00 kg hay (850g/kg D.M.) @ 8.4 MJ/kg D.M.	0.85	7.1
2.00 kg swedes (120g/kg D.M.) @ 12.8 MJ/kg D.M.	0.24	3.1
	1.09	10.2

(The arithmetic of the above is: $1 \times 0.85 \times 8.4 = 7.1$
and: $2 \times 0.12 \times 12.8 = 3.1$)

This gives an energy content a little in excess of the theoretical value and a dry matter intake below appetite. The response of a practical farmer to this situation would be to make a slight decrease in the hay supplied and to give the ewes some clean barley straw to pick over and so ensure a satisfactory gut fill.

With regard to protein, reference to the tables shows the requirement to be about 70–80 g D.C.P. per day. The above ration contains:

$$\left. \begin{array}{l} \text{Hay} \quad\ 1 \times 0.85 \times 58 = 49.3 \\ \text{Swedes} \ \ 2 \times 0.12 \times 91 = 21.8 \end{array} \right\} \quad 71.1 \text{ g/day}$$

This is slightly on the low side but as concentrate feeding will start soon the adjustment can wait until then. This feeding of concentrates will increase as the ewe's capacity for dry matter falls, and she eats less hay.

Shortly after lambing the ewe's appetite will revive and as the demands of lactation increase the ration must also be increased. The ration should contain the same ingredients as fed before lambing as far as is practicable. This is a good general rule but, should a little really good fodder such as barn-dried hay be available, feeding it from a fortnight before lambing until grass is available will be a boon to the ewes.

Example 2

Taking the same ewes a week after lambing twins, the energy demand of each ewe will have risen to 28.6 MJ/day, but the capacity for dry matter will also have risen to, say, 3.5% of body weight, which gives:

$$\frac{3.5 \times 70}{100} = 2.45 \text{ kg D.M.I. capacity per day.}$$

	D.M.I. (kg)	M.E. (MJ)
1.00 kg leafy hay (850/kg D.M.) @ 10.1 MJ/kg D.M.	0.85	8.58
2.00 kg swedes (120/kg D.M.) @ 12.8 MJ/kg D.M.	0.24	3.10
1.50 kg sheep concentrate (880/kg D.M.) @ 12.5 MJ/kg D.M.	1.32	16.50
		28.18

This is satisfactory from the point of view of energy requirement, provided the protein is sufficient. From Table 2.7 a 70 kg ewe feeding twins requires 307 g D.C.P. per day. Leafy hay contains 90 g D.C.P. per kg.

$$\begin{array}{l} \text{The hay provides} \quad\ 1 \times 0.85 \times 90 = 76.5 \text{ g} \\ \text{The swedes provide} \ \ 2 \times 0.12 \times 91 = 21.8 \text{ g} \end{array}$$

The sheep concentrate will have about 150 g/kg D.C.P.,

Therefore 1.5 kg will
contain $1.50 \times 0.88 \times 150 = 198.0$
                                          _____

                                          296.3 g of D.C.P.
                                          _____

This intake of 296 g/day D.C.P. is sufficient for all practical purposes. Some farmers might think that 1.5 kg concentrate is a lot to feed per day but this is the sort of input required if the ewe is not to 'milk off her back'.

Example 3
Finally we turn to indoor-fed lambs for an example which takes 30 kg live weight lambs for which a live weight gain of 200 g/day is anticipated.
    Each animal will have a D.M.I. capacity of about 3% of its live weight, and 3% of 30 kg is 0.9 kg. From Tables 2.6 and 2.8 such an animal will require 7.3 MJ/day and 94 g of D.C.P. Assuming a basic ration of 0.50 kg/day of clover hay and 1 kg/day of mangels, these components provide the following:

		D.M.I. (kg)	M.E. (MJ)
0.50 kg clover hay	(850g/kg D.M.) @ 9.6 MJ/kg D.M.	0.43	4.08
1.00 kg mangels (125/kg D.M.) @ 12.4 MJ/kg D.M.		0.13	1.50

This ration is too low in energy therefore a
concentrate is added:

0.30 kg concentrate	(880g/kg D.M.) @ 12.5 MJ/kg D.M.	0.27	3.30
		0.83	8.88

This ration is well within the limits for D.M.I. but in excess in respect of energy. But before we consider an adjustment we must look at the protein position:

0.50 kg clover hay	(850g/kg D.M.) @ 128g/kg D.C.P. =	54.40
1.00 kg mangels	(125g/kg D.M.) @ 58g/kg D.C.P. =	7.30

0.30 kg concentrate   (880g/kg D.M.) @ 150g/kg D.C.P. =   39.60

$$\overline{}$$

101.30

$$\overline{}$$

The above ration is within the theoretical limits for dry matter intake but supplies more energy and protein than is required for the 200 g daily gain. But, with growing sheep, this is no disadvantage so the ration would be allowed to stand. The student should also beware of the trap of trying to get exact figures. The ration could be manipulated until it met the exact theoretical requirements but the units used would be unrealistic. One does not tell a shepherd to feed 1.07 kg hay per head per day, at least not if one wishes to retain his respect!

These three examples give an indication of how farmers can work out rations. Through practice over a period of time a flockmaster is able to get quite close to theoretical requirements in making up rations without a series of abortive attempts. The main point to ensure is that there is no large discrepancy between the theoretical and the practical. It is particularly important to see that lactating ewes and growing lambs are not required to achieve too much with too little.

Further reference will be made to feeding when systems are discussed later.

Points to remember

1. Basic nutrient requirements:

    Energy
    Protein
    Water
    Minerals
    Vitamins

2. Principal minerals required:

    Calcium
    Phosphorus
    Magnesium
    Sodium
    Chlorine
    Sulphur

Copper
Cobalt
Selenium

3. Principal vitamins required:

    Vitamin A
    Vitamin D
    Vitamin E

4. Metabolisable energy:

    Metabolisable energy is the useful energy the animal extracts from its food, and is used for:

    Maintenance (including general activity)
    Production (such as lactation)
    Growth
    Pregnancy

    Energy is measured in joules and megajoules (1MJ - 1 000 000 joules) e.g. the M.E. requirement for a 50 kg dry, non-pregnant ewe is 6.8 MJ per day.

5. Protein requirements:

    Given the weight of the animal and its production situation, the protein requirement may be determined from tables, e.g. a 50kg dry, non-pregnant ewe needs 56 g per day of digestible crude protein, but 132 g per day if growing at a daily rate of 300 g live weight gain.

6. Some practical feeding points:

(a) The higher the digestibility the faster the passage of food through the gut and the more the animal can eat.
(b) Correct energy/protein balance is important especially in high production as high production implies high inputs of valuable resources. If there is imbalance wastage will occur and the potential output cannot be reached.
(c) The digestible crude protein fraction falls rapidly in grass and forage crops as they mature but the digestibility of white clover falls more slowly throughout the year than its companion grasses. This means that a pasture containing clover provides a more digestible feed over a longer period than a pure grass sward.

3 Diseases and disorders; prevention and treatment

General health considerations

Sheep, like all animals, are afflicted from time to time by a number of diseases, parasites and physiological malfunctions. Any husbandry system devised by the farmer must take note of these potential problems as disease is best tackled from the standpoint of prevention rather than cure. Those who have worked with sheep will be familiar with the old adage that 'a sheep's worst enemy is another sheep', but those who are familiar with the economics of sheep farming (at least for meat lamb production) will be aware that one of the major factors affecting profitability is the stocking rate. A low stocking rate on expensive good land will lead to poor profitability regardless of the flock's health record. The question resolves itself into one of balance: in other words, how to achieve a stocking rate appropriate to a particular system on a given farm without loss of performance due to disease. There are, of course, some diseases of sheep which are scarcely influenced by stocking rate, but many of the major diseases are.

It is important from the start that the person in charge of sheep adopts the right attitude to the health of the flock. The simplistic approach that the equation is 'sheep plus disease organism = disease – send for the vet and hope he can produce an injection that will make all well' is a bad one. There are times when 'fire brigade' action by the veterinarian is appropriate, but getting the advice of a vet before setting up the system is usually much less worrying and much more economical than leaving things to chance.

It is not the purpose of this book to list all the diseases of sheep and suggest cures and preventions. What will be done is to look at a number of cases of different types of health problems and suggest methods of avoiding or resolving them. It is hoped that from these a student will develop an approach to disease problems which will enable him to cope in the field situation.

Diseases and parasites will be looked at from a general standpoint and a start will be made with the equation, quoted above, which applies in the case of a disease like foot and mouth. This disease is highly infectious, readily spread, and in countries like Great Britain, Canada and New Zealand the sheep have no immunity; once the flock is attacked 100% infection is to be expected. On the other hand one frequently sees examples of disease in housed sheep where sheep plus an appropriate organism plus bad ventilation = pneumonia. Many disease organisms are ubiquitous, but they need suitable conditions before they can mount a successful attack.

A general aspect of disease which must be grasped is that *a disease condition need not be overt*. A typical example in sheep is sub-clinical helminthiasis or, in other words, a worm burden that is not big enough to be noticed but sufficient to slow down the growth rate of a fattening animal. It is probably true to say that sub-clinical diseases and disorders are as costly in the field of animal production as are the cases of obvious ill-health and death. In the wild the most successful animals are those best adapted to their environment. In intensive husbandry systems one can almost turn this round and say that the most economically successful animals are those whose environment has been adapted most successfully to their needs. Environment is not used here in a narrow sense. It means not only the site, aspect, soil type, climate and all the factors popularly associated with environment; it also means food supply and its quality and quantity, protection from disease by such means as inoculation and other measures, and the way in which sheep are handled and the timeliness of operations.

Factors affecting the health of sheep

We must now look at the factors which have major effects on the health of a flock in causing disease and death.

NUTRITION

One of these major factors is, of course, nutrition. If that is reduced to the extreme the animal dies. Death from starvation does not normally occur except under extensive farming conditions, such as on the high hills and semi-desert areas. On the other hand a break-down of health due to under-nutrition is by no means rare: *pregnancy toxaemia* is a typical case. The loss of young lambs due to the poor milk supply of the ewe is also frequent. The low milk supply is itself normally caused by malnutrition and this state of affairs is common in extensively run flocks.

The effect of parasite infection is also enhanced by malnutrition as is well seen in weaned lambs that carry a substantial worm burden. The above statements must not, however, lead the student into the assumption that all that is required to keep sheep healthy is to supply food *ad libitum*. On the contrary, it is normally sheep in good condition or on a rising plane of nutrition that succumb to *enterotoxaemia*, while overfat animals have a tendency to respiratory diseases.

STRESS

While malnutrition obviously applies stress to an animal there are other stress factors which play their part in triggering off disease outbreaks. Moving sheep long distances by truck often 'sparks off' *pneumonia*, while cold wet weather is associated with *hypomagnesaemia* in ewes. Sheep should always be treated in a quiet manner and not harassed by ill-trained dogs. This is particularly true of in-lamb ewes and young lambs. Finally, the farmer must always remember that one of the severest stresses a ewe has to sustain is that of lactation and everything that is reasonably possible should be done to improve her lot.

AGE

There is also an age factor in the incidence of disease. In general young animals are more susceptible to disease and parasitism than older animals. There are some diseases such as *lamb dysentery*, which do not affect adults, and parasites such as *Nematodirus* which are lethal to lambs but cause no trouble to adults.

Of the diseases and parasites from which both young and adults suffer, many are more lethal to the young than to the old. One reason for this is that the young animal is more susceptible to inclement weather than the older animal. The young lamb has a much larger surface area relative to its mass than the adult sheep and consequently suffers more readily from *hypothermia*. Apart from death due to this the animal suffering from cold has also much less resistance to disease. It will also be appreciated that the young animal has a greater food requirement relative to its mass than does the older one. There are, however, some diseases such as *scrapie* and *Johne's disease* which are not normally seen in young animals and are only manifest in older animals, although they are doubtless present sub-clinically at quite an early age.

DOSE OF INFECTION

One of the most important factors in determining whether or not an animal becomes infected with a disease or parasite is the massiveness of the dose of infection. In the case of a lamb that picks up a few stomach worm larvae, natural defences will, in all probability, contain the parasite and at the same time will build up an immunity against the worm. The same is true for various bacterially induced diseases. It is for this reason that strict hygiene is counselled for lambing pens and similar places. Under farming conditions it is quite impracticable to provide aseptic conditions but the fewer the pathogens present, the less the likelihood of disease.

IMMUNITY

The mention of immunity in the previous paragraph leads us on to another aspect of disease resistance. Whether or not an animal develops a disease when challenged by disease organisms depends largely on the degree of immunity present in the animal. This immunity may be natural or artificial, active or passive. There are also groups of animals with a natural resistance to the particular diseases with which they have come in contact. In Britain, for example, the sheep which for generations have been raised on tick-infested land have developed a hereditary resistance to such diseases as *tick-borne fever*. This is true of all areas where ticks are prevalent, and this fact has very important husbandry consequences which will be discussed later as will the provision of artificial immunity against disease, and therapeutic treatments. Husbandry techniques for preventing the spread of parasites and disease such as dipping against sheep scab, footbathing against foot rot and rotational grazing to mitigate the attacks of stomach and other worms will also be discussed later in the chapter.

ERADICATION OF PESTS, PARASITES AND DISEASE

Methods of eradicating parasites and disease can vary from a minor operation, such as draining a piece of marshy ground in an endeavour to eliminate the host snail of liver fluke, to a government slaughter policy to check foot and mouth disease, or compulsory dipping to eliminate sheep scab. It is the duty of every flockmaster to comply fully with official requirements if for no better reason than that it is in his own long-term self interest. In the long-term most disease elimination is usually a much more economical exercise than curative treatment, even in cases where a satisfactory cure is available.

Having briefly surveyed the factors affecting general health we must now look in a little more detail at some specific diseases. The number of diseases and disorders from which sheep may suffer is very large, but fortunately these do not occur all at once, or all in the same area. There are, however, a number which occur fairly regularly, and we will confine our attention to these.

Bacterial diseases

PROTECTIVE MEASURES

One of the main defences an animal can mobilise against a pathogenic organism is the production of specific antibodies. If an animal has previously been challenged by a disease and overcome it, it will have developed some immunity against the disease. This is known as *active immunity*. For many diseases it is possible to produce such an immunity by artificial means. An attenuated or killed organism is introduced into the animal, usually by means of an injection. This does not set up the disease in an active form but stimulates the animal to produce *antibodies*. This treatment is known as *vaccination*. This animal when challenged by the virulent disease organism is then in a position to make an immediate response. It should be noted that for a number of diseases more than one vaccination is necessary, with a lapse of some weeks between the two injections before the treatment becomes effective. Another point of importance is that few, if any, vaccines confer lifetime immunity from disease and inoculation has to be carried out at regular intervals.

Another protective measure that can be taken against some diseases is the use of *serum*. This produces *passive immunity*. In other words it does not stimulate the animal to produce its own antibodies. The serum works because it contains antibodies which have been produced in some other animal, such as via the blood of a horse. It gives only temporary protection because the antibodies, like all extraneous material introduced into the blood stream of an animal, are ultimately destroyed. Their usual active period is about 10 to 14 days. A serum is used where temporary protection is needed, as in the case of *lamb dysentery* where the organism endangers the lamb only for the first fortnight or so of its life. In other cases serum can be used in conjunction with a vaccine to give cover until such time as the vaccine has activated the animal into producing its own antibodies. An example of this is the use of both vaccine and serum in an outbreak of *pulpy kidney disease* in lambs. The major drugs used in the case of diseases of bacterial origin are antibiotics for which, in many countries including Britain, a veterinary prescription is required.

Sheep are amongst the easiest domestic animals to protect from disease by such prophylactic treatments as the use of vaccines and sera, and consequently these are widely used. This statement may give the impression to a student who has a limited knowledge of physiology that antibodies are the only lines of natural defence an animal has. Such is not the case. The subject is very complicated and interested students are advised to read further in books specialising in veterinary immunology.

CLOSTRIDIAL DISEASES

To turn to specific organisms, the group of bacteria which cause most trouble amongst sheep are the *clostridia*. Some members of the group are the causative organisms of diseases in other animals, including man. Two relatively common diseases of man attributable to clostridia are *tetanus* and *gas gangrene*. One of the chief villains causing sheep diseases is *Clostridium welchii*. This organism exists in a number of different types. *C. welchii type D* causes a disease which in lambs is known as *pulpy kidney*, and in older sheep as *enterotoxaemia*. *C. welchii type B* causes *lamb dysentery*. These diseases are both killers and can be the cause of grievous loss.

The lamb dysentery organism persists in the ground for long periods and it is endemic in many areas of Britain. In some areas it is unknown, but in areas where it is present lambs should be protected. This can be done in one of two ways: either the mother can be inoculated with vaccine about tupping time and again just before lambing, or the lamb can be given an injection of serum within twenty-four hours of birth. The surest method is to inject the lamb with serum. However, this is difficult under extensive systems. The vaccination of the ewe promotes antibody formation, these antibodies being passed on to the lamb in the colostrum, but the efficiency of this method depends on the lamb receiving an adequate feed of colostrum. Under British conditions the tendency is for farmers to make use of combined vaccines which besides giving protection against enterotoxaemia and lamb dysentery, also cover *struck* (*C. welchii type C*), *braxy* (*C. septique*), *blackleg* (*C. chauvoei*), *black disease* (*C. oedematiens*) and *tetanus* (*C. tetani*).

As with all diseases there are contributory factors which influence the outbreaks of clostridial diseases. Enterotoxaemia and pulpy kidney disease normally strike sheep which are on a rising plane of nutrition and as *C. welchii type D* is usually present in the gut of the sheep or lambs the disease can strike the whole flock quite suddenly. The first indication of trouble in a non-protected meat lamb flock is to find the best lamb dead one morning. The contributory factor involved in tetanus is a wound whereby the organism has gained access. This commonly happens during tailing, castration or shearing.

Black disease is brought about by an invasion of the liver by immature liver fluke which damages the liver tissue which is in turn attacked by *C. oedematiens*. Wounds can also become infected by other members of the clostridia such as *C. chauvoei* and *C. septique*. This includes infection of the womb at parturition. A number of the above organisms such as *C. tetani* are spore-forming and can contaminate land and buildings for long periods. It is therefore imperative that strict hygiene should be observed in all sheep-handling operations and in all buildings where sheep are held. While the clostridial diseases are the most important bacterial diseases which infect sheep there are some others which cause substantial problems, chief of which is *foot rot*.

FOOT ROT

Foot rot and *liver fluke disease* (see later in this chapter) have caused havoc with domestic sheep probably ever since neolithic man brought them out of the dry lands of the Middle East to the high rainfall areas and lush pastures of Western Europe. Dealing with foot rot in detail will be deferred until the husbandry section in Chapter 5, but there are a number of points concerning the causative organism which it is appropriate to make here:

- The organism is *Fusiformis nodosus* and it attacks only sheep and goats.
- It cannot survive away from the animal for more than about a fortnight.
- It cannot attack a dry undamaged foot.
- It cannot attack when the weather is cold, but it can survive on the foot at any normal temperatures.

Under suitable conditions the disease can spread very rapidly, with the majority of the flock becoming lame. Wet, warm conditions with the sheep grazing a lush pasture are ideal for the spread of the disease. Foot rot is not a killing disease but the discomfort and inability to move readily soon takes the sufferer down in condition. There are other organisms that can cause lameness such as *Fusiformis necrophorus*, together with *Corynebacterium pyogenes* and various streptococcal and staphylococcal organisms.

SEPTIC CONDITIONS

Joint ill is associated with various organisms such as streptococci, coliform bacteria and corynebacteria and is a common problem under intensive conditions. The disease affects the joints of young lambs, the organisms entering the blood stream via the umbilical cord. This problem, together with scouring caused by *Escherichia coli*, can best be controlled by practising strict hygiene in the lambing quarters. A similar condition to joint ill arises on tick-infested farms and is known as *tick pyaemia*. This is caused by bacteria transmitted to the lamb by tick bite, various streptoccocal and staphylococcal organisms such as *Staphylococcus aureus* being implicated.

Mention has been made of clostridial organisms which attack man as well as sheep. In addition to these there is a further bacterium which can readily prove fatal. It is *B. anthracis* which causes *anthrax*. In Britain this is a notifiable disease where it usually occurs as a result of contamination of foodstuffs by spores associated with imported hides, fertilisers and similar merchandise. In some areas of the world where it is endemic the sheep are highly resistant to the disease. The spores of the organism are very resistant to weather and other conditions and can lie dormant in the soil for decades. In Britain the disease is much more common in cattle than in sheep and the main point of mentioning it is that the carcase of an anthrax casualty should never be opened up. The whole bloodstream of the animal has usually been invaded by the organism and as it is spore-forming a very dangerous situation can arise from blood-letting.

WASTING DISEASES

A number of sheep diseases are typified by the sheep pining and wasting away. The specific disease of *pine* has already been discussed in Chapter 2, but this is caused not by bacteria but cobalt deficiency. Although it does not afflict sheep in large numbers a further wasting disease which should, if possible, be eradicated is *Johne's disease*. This is caused by *Mycobacterium paratuberculosis*. Cattle also suffer from the disease and it is possible that there is cross-infection. The disease develops slowly and the animal can be infected and infective for many months before clinical signs are seen. The only positive diagnosis is by post mortem examination. The source of infection is the faeces of infected animals. The organism can live for long periods away from its host, and ponds and other stagnant waters are amongst the reservoirs of infection and should be fenced away from stock.

This type of disease underlines one of the most important aspects of sheep husbandry, namely that the unit with which one is concerned is the *flock* rather than the individual sheep. Individuals which, over a period of time are seen to be unthrifty in an otherwise healthy flock should be disposed of and where possible a veterinary post mortem should be held. Carrier animals of any disease from Johne's to persistent foot rot need to be eliminated, otherwise the rest of the flock will be constantly at risk.

Finally, as with all domestic mammals, the sheep is subject to *mastitis*. This condition is caused by various organisms, the commonest being *Staphylococcus aureus*. Prevention of the complaint is difficult, but ewes should be examined when the breeding flock is being made up and ewes showing unnatural thickening of the tissue of the udder or teat canals should be rejected. The condition is more common amongst housed ewes and those in lambing yards than for those on free range. Veterinary treatment is by antibiotics.

A number of diseases of bacterial origin which have always caused some problems in the past are tending now to cause more, especially where intensive husbandry is practised and the sheep are housed.

COCCIDIOSIS

This is a suitable place to mention *coccidiosis* although it is caused by a protozoan parasite, not a bacterium. Different species of coccidia inhabit the alimentary tracts of various animals and birds without causing adverse symptoms, but under crowded conditions such as lambing yards lambs can become heavily infested and exhibit signs of disease. The symptoms are general unthriftiness, a tucked-up appearance and scouring. The excrement is often bloodstained.

Good hygiene is essential to prevent this disease, but when an outbreak does occur veterinary advice should be sought in order to bring the most appropriate coccidiostat into immediate use. These drugs are administered orally.

Viral diseases

FOOT AND MOUTH DISEASE

The viral disease which is most dreaded amongst livestock farmers in Britain, North America, and the Antipodes is *foot and mouth*. The disease occurs in a number of types, and although there are vaccines, controlling the disease by vaccination is difficult. The above mentioned countries rely on quarantine and slaughter to keep the disease at bay. Foot and mouth disease is not a killer where it is endemic, but it is highly infective to all cloven-hooved animals. It is very debilitating and can give rise to abortion. In countries such as Britain where there is a slaughter policy, it is a notifiable disease. This means that the flockmaster is obliged to report a suspicion of the disease to the Divisional Veterinary Officer or the police. The symptoms of the disease in sheep are not as obvious as in cattle. Salivation and dribbling from the mouth is not common. Sheep tend to become lame with blisters round the coronet and also show mouth lesions.

ABORTION

Enzootic abortion or *kebbing* is fairly widespread in Britain and several other countries. This virus infection is spread at lambing time, the ewes which become infected aborting during their next pregnancy. The abortions are generally late in pregnancy and are the only overt sign that the ewe has the disease. There is an effective vaccine but it is much more satisfactory to keep clear of the disease by not buying in females, or by being very selective in their purchase. In addition to the above disease there are various other organisms causing abortion in ewes. A rickettsia, *Coxiella burneti*, which is transmitted by ticks and causes *Q Fever* is one; *Campylobacter foetus*, which causes *campylobacter foetus disease*, is another. Abortions are also caused by bacteria such as *Salmonella abortus*.

Amongst other diseases which can be responsible for abortion are: *louping ill, tick-borne fever, pregnancy toxaemia, Rift Valley fever* and *blue tongue*. Indeed anything which causes the ewe to run a high temperature or suffer other forms of stress is conducive to abortion. In any healthy and well-conducted flock there are always a few abortions and still-births at lambing time. These are mainly of mechanical origin – caused, for instance, by butting of other animals or by being knocked down – but if at any time the shepherd feels that the number is above normal or that the circumstances are in any way strange a veterinary examination of the flock should be made. The best method of controlling abortion is to establish a closed flock where the only new animals introduced from outside are bought-in rams. If this is not possible the watch-word when buying is circumspection — identify a clean source of supply of young females and stick to that source.

SCRAPIE

Scrapie has a long incubation period and is a disease in which the animal becomes itchy, excitable and nervous. The sufferers tend to rub themselves against any suitable objects, such as fencing posts, for long periods. They bite and nibble at their own bodies and in general give every impression of being thoroughly uncomfortable. As the disease progresses the sheep become thinner and thinner until death intervenes. The disease is world-wide in distribution and it has attracted substantial research but to date there is no cure. It has been artificially transferred from one sheep to another but the mode of transfer in nature is still under debate although it appears that the breeding ewe is the likely culprit. The incubation period is long — eighteen months or more — which is one reason why research has been so protracted. The only preventive measure that can be taken is exercising care in the selection of foundation stock, and culling female lines of scrapie-affected ewes. There appears to be an hereditary factor active in the resistance of some sheep to this disease.

MAEDI-VISNA

Maedi-visna is a progressive pulmonary disease which is caused by a virus and transmitted from ewe to lamb in suckling and between sheep in contact. Like scrapie it has a long incubation period and is fatal. The disease can be detected by laboratory examination of blood samples but there is to date no preventive treatment, no cure and no alternative to slaughtering infected animals. The disease occurs in the Northern Hemisphere but has not been an important problem until recently, except in Iceland. The disease could prove more and more of a problem in Britain as more ewes are winter-housed.

Parasites of the sheep

Repeated references have been made to parasites as vectors of disease, and we now look briefly at some that are of importance in Britain. The lessons they provide are also relevant to similar situations in other countries.

Parasites are divided into two classes – ectoparasites which attack the outside of the host, and endoparasites which live at least part of their life cycle within the body of the host. Different species of parasites have very different life styles. The life cycle may be simple, as in the case of stomach worms such as *Haemonchus*, where eggs from the adult female worms pass to the ground in the faeces, the eggs hatch, and pass through a number of larval stages to become actively infective (see Fig. 3.1). These are picked up by the sheep in grazing and the cycle restarts. The *lung worms* which cause *parasitic pneumonia* or *husk*, behave similarly.

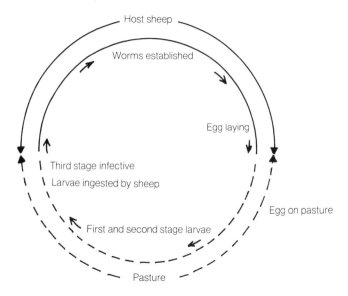

Host sheep

Worms established

Egg laying

Third stage infective

Larvae ingested by sheep

Egg on pasture

First and second stage larvae

Pasture

Fig. 3.1 Example of a parasite with a simple life cycle – *Haemonchus* sp. (stomach worm).

A parasite of sheep which has caused much trouble in some areas of Britain over the past few decades is *Nematodirus*, a roundworm which attacks lambs in spring and early summer and inhabits the small intestine. This worm has a lifecycle which occupies a year, as opposed to most worm parasites for which the cycle may be completed in a few weeks under suitable weather conditions. On the other hand, the liver-fluke *Fasciola hepatica* follows a very complicated cycle (see Fig. 3.2). The eggs of this animal pass from the bile duct of the sheep where the adult flukes are established, to the

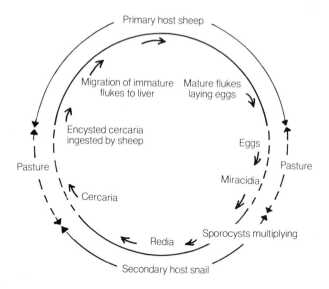

Fig. 3.2 Parasite with a complex life cycle — *Fasciola* sp. (liver fluke).

pasture. The eggs hatch and develop into first stage larvae called *miracidia* which in turn are picked up by a mud snail called *Limnaea truncatula*. Within the snail they undergo changes finally leaving to encyst on the grass. These cysts are extremely resistant to adverse weather and many survive to be eaten by other sheep. On entering the sheep the *cercaria*, as they are called, penetrate the intestinal wall and migrate to the liver. As they travel through the liver to the bile duct they increase in size and cause serious damage to the liver tissue. This damage exposes the sheep to the danger of *black disease*. The cycle takes some months to complete and as the activity of the snail is seasonal the manifestations of the disease wax and wane. In Britain, the snail's season of activity is from May to September. It can be seen from the above that if the mud snail were eliminated the disease would disappear from the sheep, and this is one of the lines of attack developed against this parasite.

A detailed knowledge of life cycles is of paramount importance in dealing with parasites. In making war on them the farmer has, as in other warlike activities, to strike the enemy at his weakest point. In the case of the liver fluke the snail can be attacked by drainage and in some cases by treating snail-infected areas with copper sulphate. The depredations of the roundworm Nematodirus can be prevented by grazing lambs only on clean reseeds, while treating infected animals by drenching with antihelminthics.

There is a seasonality in the activity of many parasites and the tick, *Ixodes ricinus*, is no exception. The tick does not reside permanently on the sheep but attaches itself to its host for a period of a few days in the spring and early summer and again in the autumn in order to engorge itself with blood. It takes the tick three years from hatching to becoming adult, passing through a larval and nymphal stage. The blood meals are required before each larval change and before the adult female can commence egg laying. The eggs are laid on rough vegetation and not on the sheep. It will be seen that this mode of life raises problems of control, as getting rid of the ticks needs weekly dippings or similar treatment over a period of months. This problem is further complicated by the fact that the tick has alternative hosts. Its blood requirements can be supplied by any warm-blooded animal such as deer, hares, rabbits and ground-nesting birds. If this were not the case and the tick were confined solely to the sheep, one line of attack would be to remove the sheep from a given area for a suitable length of time, clear the sheep by dipping, and starve into submission those ticks left on the ground. During the period of greatest multiplication of the tick population, dipping goes some way to mitigating their worst depredations but it cannot control tick-borne diseases.

Unlike the tick, organisms such as the *ked* (*Melophagus ovinus*) which is a wingless insect, the *sheep scab mites* (*Psoroptes communis* and *Sarcoptes scabeii*) and various biting and sucking lice which live permanently on the sheep can be fully controlled by suitable dipping. Dipping will be discussed more fully later in the book, as will other preventive measures to be adopted against other parasites, amongst them the sheep blow fly.

Physiological disturbances and poisoning in sheep

Having briefly examined a number of diseases caused by living organisms we must now look at physiological upsets. One such that has caused considerable trouble over the past few decades, especially under intensive grazing conditions, is *hypomagnesaemia*, known also as *grass staggers* or *magnesium tetany*. It is caused by a sudden drop in the blood magnesium of the sheep which it is unable to make good from its body reserves. It commonly occurs in lactating ewes but is not confined to them.

There are usually a number of contributory factors. Animals

grazing lush spring grass recently fertilised with nitrogen and, more importantly, potassium are at risk, and the risks seem to be worst under adverse weather conditions or when the ewes have been transported by vehicles or moved distances by dogs, i.e. when they are under stress. If the animal is noticed in time the condition can be cured with a massive subcutaneous injection of magnesium sulphate solution — say 100 ml. The disease is characterised in its early stages by a twitching of the muscles and excitability, but it is sudden in its onset and death rapidly ensues. The usual intimation that anything is wrong is to find a ewe lying dead.

The reader may well have been struck by the fact that until now little has been said about the signs and symptoms of diseases in sheep. This omission has been deliberate and for two reasons. Firstly the book does not set out to show how to be a home veterinarian. Secondly, accurate diagnosis in most sheep diseases is difficult, and in many very difficult. Indeed, in a large number of cases the well-informed veterinarian requires the services of a pathological laboratory in order to make a firm diagnosis. The over-riding object of this chapter is to emphasise the importance of disease prevention and the provision of an environment which facilitates the economical production of healthy sheep. If this is achieved the treatment of any diseases which still arise is best handled by a vet.

This is an opportune place to pay tribute to those to whom the sheep industry, at least in Britain, owes its biggest debt — namely the veterinary investigation officers. Every sheep producer needs the backing of such experts. Despite this, however, there are circumstances when the shepherd must take first aid action, and our case of hypomagnesaemia is one. Sudden illness in sheep, in which death is only delayed by hours, is usually caused by either enterotoxaemia or hypomagnesaemia. These problems usually occur at lambing time. Another condition often associated with hypomagnesaemia is *hypocalcaemia*, or *milk fever*. This condition, occurring singly, does not bring about sudden death; the animal takes a number of hours before passing quietly away. It is not possible for even a veterinarian to disentangle the complicated situation of these diseases occurring concurrently in the time available so nothing is lost by attempting first aid. The treatment adopted by many for any ewe found prostrate is a hypodermic injection of a large dose of enterotoxaemia serum followed by an equally massive dose of a mixture of calcium borogluconate and magnesium sulphate, say 100 ml. In the author's experience it is surprising how often, in a triumph of empiricism over the scientific, a ewe so treated has eventually got up and walked away.

Whilst there is little that can be done to prevent milk fever other than to ensure adequate calcium in the diet, hypomagnesaemia can be readily controlled in low ground sheep. The preventive measure is to feed 15 g (½ oz) calcined magnesite per 450g (1 lb) concentrate feed per day. For ewes running out of doors the magnesite should be introduced into the concentrate ration about a week before lambing or for ewes lambed inside a week before turning out to grass. Alternatively, magnesium 'bullets' can now be obtained. These are lodged in the reticulum of the sheep where they make magnesium slowly available.

A further metabolic upset which has already been mentioned is *pregnancy toxaemia*, or *twin lamb disease*. This usually occurs in ewes carrying twin lambs and which are undernourished due to low intake or inferior food. It also occurs in ewes which have been allowed to become over-fat. Additionally it can happen to ewes which have been adequately fed and are in good condition, but which have suddenly been deprived of food, such as ewes which have been snowed in. The disease occurs in late pregnancy and is often fatal. If the ewe lambs she often recovers, but quite commonly the lambs are still-born. The condition appears to arise from the ewe being unable to ingest enough food to satisfy the requirements of herself and her unborn lambs. She then draws on her body reserves: something goes wrong with the catabolic process, blood glucose falls to a low level and ketone bodies are present in the blood. Curative measures are rarely very satisfactory, but a high-quality energy food, such as flaked maize, together with the administration of glycerine by mouth may prove helpful. The message once again is that ewes should be kept in good, well-fleshed, firm condition but not allowed to get fat. Large quantities of foods such as turnips, swedes, mangels or poor quality silage should not be fed during the last two months of pregnancy.

A number of metabolic disorders were covered in the previous chapter in conjunction with mineral and vitamin deficiencies when discussing nutrition. This, together with the conditions described in this section, will have given the reader insight into the metabolic problems a flockmaster must guard against.

Poisoning amongst sheep

POISONING CAUSED BY PLANTS

In most areas where sheep are kept there are plants which will, if eaten in quantity, cause upsets and in some cases death. Examples are, in Britain the yew, rhododenron and laurel (these are especially dangerous when the ground is snow-covered), in Eastern Canada the laurel, and in New Zealand ragwort. If, at any time, an outbreak of what appears to be poisoning occurs, veterinary assistance should be called in immediately. The best way of dealing with known plant poisons is either to eliminate the plant or to fence stock away from it. It is up to each farmer to familiarise himself with the poisonous plants of his area and to take proper precautions.

Whilst poisoning amongst sheep is not common, ill health and loss can result from the injudicious feeding of certain crop plants. This is true in respect of the brassicas and rape and kale poisoning are not unduly rare. The condition occurs when stock are fenced onto pure stands of the crop without recource to other foods such as hay, and can be readily noticed as the poisoning gives rise to *haemoglobinuria* (the urine becoming red from the haemoglobin present). Sheep on these crops should either receive hay, or be run back onto pasture for a large part of each day.

Problems can also arise from plant substances setting up undesirable reactions in the animal. A typical example is a disease known in Scotland as *yellows*; bog asphodel is one plant which causes this effect. What happens is that a photosensitive substance, produced by the plant, accumulates in the blood and is acted on by sunlight. White-faced sheep — especially lambs — suffer, exhibiting swollen ears and swellings of the face which cause acute discomfort but are seldom fatal. Some strains of subterranean clover have been indicted in Australia as the cause of infertility in ewes, the high oestrogen levels in these plants being responsible, while in New Zealand the ingestion of fungi can give rise to serious conditions of facial eczema.

MAN-MADE POISONS

The salts of the heavy metals are potent killers and this is particularly true of *lead*. Lead salts have the unfortunate property of being sweet to the taste and therefore attractive to animals. Lead-based paints should never be used on surfaces which can be reached by animals, and paint containers should be carefully disposed of.

Copper has been mentioned under nutrition in Chapter 2 and is one of the more dangerous mineral poisons. It is likely to cause problems because it is fairly accessible and is cumulative. *Arsenic* was a common culprit when used in sheep dips but today its importance has declined. The most likely sources of poisons today under intensive farming conditions are *herbicides, insecticides* (not forgetting some sheep dips) and *fungicides.*

The most important factor in the prevention of poisoning is arranging the safe disposal of containers. Where any treatment of crops to be grazed, such as top dressing or spraying, has been undertaken, stock should not be returned to the grazing before a suitable lapse of time and until there has been a reasonable fall of rain. The author, however, has had the experience of losing lambs in spite of these precautions. A field which had been treated with basic slag and well rained upon was used to pasture ewes with young lambs at foot. Three days later lambs started scouring and some died. Fortunately the lambs were quickly referred to the veterinary investigation officer. Slag poisoning was immediately diagnosed, the sheep were removed from the field and the trouble ended. What had happened was that the lambs of about two to three weeks of age had found the little piles of slag where the paper container bags had been burned and then sampled their find in a too enthusiastic way. This little tale is related not to suggest that large numbers of lambs are lost from basic slag poisoning or that it is of major significance, but to illustrate that when a health problem arises every aspect of the case should be examined. It should not be assumed that because a precaution has been taken against a certain danger the precaution has been effective.

The second point to be made is that when such a problem arises professional advice should be sought immediately. The fact that only a handful of lambs died might seem to be making a mountain out of a molehill, but had the diagnosis been a few days later, and the animals left with undisturbed access to the slag the direct losses would have been much greater. The indirect losses due to the setbacks sustained later by the lambs which did not die would also have been considerably more.

Whilst it is not possible to mention every kind of poison to which sheep might be at risk, it is hoped that the above examples will alert farmers and students to the fact that they should try to remain aware of the various problems of this kind that arise from time to time.

Hereditary abnormalities

Hereditary abnormalities are not as common in sheep as they are, for instance, in dogs and swine. Two of the commonest defects are undershot and overshot jaw. This is where the lower jaw is of such a length that the incisor teeth do not meet the dental pad properly (see also Chapter 2). This condition naturally impairs the grazing ability of the sufferer and could be recessive in character. The best line of action, therefore, is to eliminate both sire and dam from the breeding flock. *Daft lamb disease*, where the lamb carries its head in an unnatural position and is unable to suckle, is another condition. This is seen principally in the Border Leicester breed. Here again the appropriate action is to eliminate both parents. *Cud spilling* is another problem which afflicts certain sheep, mostly from breeds with pronounced 'Roman' noses.

In addition to physical abnormalities and physiological malfunctions which are due to genetic factors, heredity plays an important part in determining how an animal reacts to the challenge of disease-producing organisms, shortages of particular elements in the diet and to its response to its physical environment. For example, some breeds such as the Merino are more susceptible to foot rot than, say, the Romney Marsh.

Conclusion

The student whose first introduction to sheep is this book may well feel alarmed at all the misfortunes that can befall the animal. On the other hand, he may be so encouraged by the list of preventive measures that are effective against diseases to think that a vast veterinary umbrella can be erected over the world of sheep whereby all are rendered safe.

When husbandry systems are discussed later an effort will be made to show that in a well-organised system neither disaster from disease nor economic collapse from over-expenditure on health preservation need be the outcome of a sheep-keeping enterprise. Disregard of any form of preventive measures, even on an extensive system, usually leads to failure, but a sheep that has received all the preventive inoculations, all the drenches, all the food additives, has been dipped and sprayed with all the insecticides and acaricides that persons knowledgable in the appropriate subjects have deemed desirable, will leave little profit. The art of successful stock-raising lies in striking a balance between the theoretically desirable and the economically practicable. This means that in a country such as Britain, the sheep flock should receive as routine:

- Preventive inoculation against clostridial diseases
- Spraying or dipping against ectoparasites
- Some drenching against stomach worms
- Regular foot rot treatment

Treatments other than these should wait on veterinary consultation and a diagnosis of a specific condition in need of a particular treatment.

Points to remember

1. Causes of disease:

(a) Living organisms such as bacteria, viruses, protozoa, parasites and fungi.
(b) Physiological disorders induced by malnutrition, dietary imbalance, mineral or vitamin deficiencies, and trauma.

2. Factors affecting the outbreak of disease:

Nutritional status.
Stress.
Age.
Massiveness of dose of infection.
Degree of immunity.
Husbandry techniques to prevent the spread of disease and parasites.

3. Eradication of parasites and disease:

Long-term prevention is usually much more economical than cure.
Methods range from dipping to prevent sheep scab, footbathing against foot rot, rotational grazing to mitigate attacks of stomach and other worms, draining marshy ground to eliminate the host snail of liver fluke, to government slaughter policies against foot and mouth disease.

4. Routine preventive measures necessary in Britain:

Inoculation against clostridial diseases.

Dipping or spraying against ectoparasites.
Drenching against stomach worms.
Footbathing against foot rot.
Other treatments should wait on veterinary advice.

5. In general:

(a) In the wild the most successful animals are those best adapted to their environment. In intensive husbandry systems the most economically successful animals are those whose environment has been most successfully adapted to their needs.

(b) Environment means not only site, aspect, soil type, climate, food supply and its quality and quantity but also protection from disease by inoculation and hygienic measures, handling and timeliness of operations.

(c) The art of successful stock raising lies in striking a balance between what is theoretically desirable on the one hand and economically practicable on the other.

4 The domestication of sheep and the evolution of modern breeds

The domestication of sheep

The first animal to be domesticated, by early Stone Age man, was the dog. It would not have been a premeditated, conscious action but what one could call almost a symbiotic process. Wild dogs and wolves have a tendency to scavenge around the habitations of man and what probably happened was that the dogs became more and more fearless of man and he found the animals more and more acceptable. The adoption of stray puppies as pets was probably a contributory factor in the process, as the practice of attempting to rear wild animals as pets is universal.

Over a period of time the dog became thoroughly integrated into the life of the family group or tribe. Dogs would help in the hunt, give warning of large predators such as wolves, and also of men foreign to the tribe. Men and dogs probably followed the flocks of sheep and goats as they moved from one grazing area to another, in much the same way as the Eskimo and American Indian followed the caribou, the sheep providing meat for food and skins for clothing. Many wolves and wild dogs exhibit the trait of heading off and ambushing prey they are pursuing, and this behaviour would doubtless be exhibited by Stone Age man's canine associates. As man extended his influence over the dog and the rapport between the two species became stronger, a situation eventually arose whereby man was so far in control of the dog as to be able, with the animal's help, to round up the sheep and persuade the dog to engage in a holding operation rather than in an attack. From that day on man became a herdsman. When or where this evolution from a hunter and gatherer to a herdsman took place is unknown. It is quite possible that there was more than one centre of domestication and more than one period in which it occurred. The probable area and time is thought by archaeologists to be the highlands of Anatolia and Iran, in the mesolithic period of some twelve thousand years ago. By the time the neolithic revolution was in full swing some men and their flocks and herds (by now they had acquired cattle) had started moving out of the Middle East along the Mediterranean into Western Europe, and by about 3000 B.C. they had arrived in Britain.

How long it took the primitive stockman to acquire the knowledge about his animals we now take for granted will never be known. When did he first realise that a flock or herd required relatively few males and that it was advantageous to himself to kill off males in preference to females? When did he learn to compute the gestation periods of various animals? Somewhere along man's evolutionary road he discovered that sheep could make further contributions to his well-being other than meat and skins. He discovered that the soft undercoat or wool of the sheep could be plucked from the living animal, spun into thread and woven into cloth. He found that he could hand-milk the ewe to provide himself with another source of food. In due course he discovered the process of cheese-making which gave him a concentrated protein food which could be stored for an appreciable period. He also found out how to make candles from tallow which, in the higher latitudes, was important in extending the time of activity spent on craft work during the long winter nights. Finally, within the last century came the discovery of effective refrigeration on a factory scale, which enabled meat to be kept over long periods, and transported from one side of the world to the other.

As sheep husbandry had its origins in the Middle East it is reasonable to suppose that the first sheep domesticated was the Urial (*Ovis vignei*) and was probably the first to come under the control of man. As the thrust of the neolithic revolution expanded westwards along the Mediterranean, blood of the Mouflon (*O. musimom*) would also become incorporated into the flocks of the pioneer herdsmen. Having arrived in Western Europe, we have come to the area from which the majority of sheep breeds used in the commercial wool and mutton trade of the world have evolved.

The evolution of modern sheep types

Two premier sheep countries of Europe are Spain and Britain. Not only do they carry quite large sheep stocks at the present day, but they also made a major contribution to sheep population of the world. There is considerable debate as to what the first sheep in Britain was like. In all probability it was an animal with the characteristics of the Soay — small in size, brown in colour and short-tailed. Accepting that a Soay-like sheep was the first introduced into England by the Neolithic people, the successive waves of incomers of the Bronze and Iron Ages would have brought in other types, and by Roman times there were other much larger sheep with white fleeces present. Sheep farming appears to have been an important activity in England during the Roman era. This importance re-emerged in the medieval period to become the most important section of livestock husbandry and the backbone of the country's trade. The twelfth century saw a great expansion of wool production — in many cases spearheaded by the monastic orders, such as the Cistercians of Rievaulx in Yorkshire, and their brothers of Melrose in the Tweed Valley. Landowners, other than the Church, became active and sheep farming all over England and Southern Scotland boomed. It was a time of the golden fleece and the fortunes amassed by merchants are still proclaimed by the 'wool churches' from Northleach to Lavenham. The Lord Chancellor sat on the woolsack, and the Middle Ages in England saw an importance attached to sheep that was not to be repeated in any country until the nineteenth century in Australia. The sheep on which this wool trade depended was a longwool of the Cotswold type but it is much too early to talk of *breeds* which were an eighteenth and nineteenth century phenomenon.

Throughout the period from the Norman conquest, and probably much earlier, an important part of the sheep's contribution to prosperity was not only the production of wool but also helping to maintain the fertility of arable land. In the manorial period sheep were grazed on the commons during the day under the care of a shepherd who returned them to the arable land for the night so that their droppings could put some badly needed heart into the corn land. From the fourteenth century with its decline of the three field system and the increase in enclosures, more and more sheep would have been kept on grassland. These sheep were long-legged, white-faced and fleeced, with poor mutton conformation, and while there would be variations from one area to another, no such thing as breeds had evolved. These, then, were the sheep that came down to the eighteenth century and the agricultural improvers.

The Agricultural Revolution and modern breeds

The upsurge of invention and experimentation, known as the agricultural revolution, had its origins in the Low Countries in the seventeenth century. New ideas and new crops made their way from there to England in the late seventeenth and early eighteenth centuries. This formed a gestation period which in due course, gave birth to the Agricultural Revolution, which occurred during the period 1750 to 1880. Within this period many farmers were working on the improvement of livestock as well as experimenting with new crops and new farming systems.

ROBERT BAKEWELL

One of the experimenters who achieved renown as a livestock improver was *Robert Bakewell* (1725–95) of Dishley Grange in Leicestershire. He worked with both cattle and sheep. His work with Longhorn cattle came to nothing, but his success with Leicester sheep reverberates through to the present day. What Bakewell did with the indigenous Leicestershire sheep was to diminish their size of frame, improve their growth rate, and lighten the fleece weight. He did this through selecting by eye and touch what he considered to be the best sheep, and he probably ranged the countryside to find such animals. He bred these selected sheep together, and re-selected from their progeny. He also leased out selected rams to other farmers and, in effect, ran a progeny testing scheme. This practice of hiring out rams was also common in Lincolnshire where a number of breeders were working on the improvement of the Lincoln breed.

Having produced animals which met his requirements, Bakewell inbred them to fix their type but the 'New Leicester', as Bakewell's sheep were called, did not meet with universal acclaim. Many people considered them too fat and their flesh insipid. It should be remembered that fat did not present the same drawback in a carcase of those days as it does today. To people who had no paraffin wax, let alone kerosene or electric light, tallow was also a very useful commodity. The modern Leicester sheep, however, is a low set blocky animal, the ewes having live weights approaching 100 kg; good quality ewes clip about 5 kg per fleece. The reason for going into detail about the Leicester and Lincoln breeds and their origins in a book which seeks to concentrate on the general aspects of sheep-keeping rather than the detail, is that the Leicester and the Lincoln have had a great influence on sheep breeding over large areas of the world.

It seems that besides direct derivatives, such as the Border Leicester and the Bluefaced Leicester, Bakewell's Leicester 'blood' influenced all the British longwool breeds. These in their turn influenced sheep husbandry in all corners of the world. The mutton qualities of Bakewell's sheep were questionable but there were other eighteenth century improvers who directed their attention to carcase quality.

JOHN ELLMAN AND JONAS WEBB: THE SOUTHDOWN AND SUFFOLK

The leading figure in carcase improvement was John Ellman (1753–1832) of Glynde in Sussex. Ellman is said to have sought the advice of butchers in his search for the ideal mutton sheep. He started with the speckled-faced, short-woolled, indigenous sheep of the Sussex Downs and as far as is known did not introduce 'blood' from outside. The work of Ellman was furthered in the early nineteenth century by Jonas Webb of Babraham in Cambridgeshire. The present day Southdown which is the outcome of their work is a short-legged, small, blocky sheep with heavy hind quarters. The ewes weigh 50 kg plus, and the clip is about 1½ kg per fleece.

Southdown 'blood' has been used in the formation of all other Down breeds. Today the leading Down breed in Britain based on numbers of rams in use, is the Suffolk. This breed evolved from a cross between the Southdown and the old Norfolk Horned – another heath breed. While early maturity and excellent mutton conformation and quality was attained, there were some aspects of performance which were lost sight of, the most important being prolificacy and milking ability although the Suffolk is less deficient in these respects than most other Downs. The result of this state of affairs is that the Down breeds are kept only for the production of replacements and of pure-bred rams to be used for fathering meat lambs from ewes of other breeds and crosses.

Traditional systems

What has been said about the improvement of low ground sheep by large scale farmers during the eighteenth and nineteenth centuries should not obscure the fact that the small farmers and crofters of those days would continue to use sheep as triple-purpose animals without giving much thought to their improvement. The large-scale milking of ewes for cheese-making had been carried out in Essex and Somerset for some centuries, but these enterprises died out towards the end of the eighteenth century. The cottagers and crofters, however, continued with ewe milking in some areas until into the present century.

In many cases in the mountainous areas of the country, especially in Wales and the Scottish Highlands, the mode of operation was the *Shieling System* whereby the cattle and sheep were sent up to the mountain pastures during the summer to be tended by the women and children, who returned with them to the lowland homestead in autumn. Similar transhumance systems were practised in other areas of Europe and the Near East. Over the past 250 years, however, a marked change has come over the hill farming of Britain. In some areas the change was sudden and dramatic, as where 'the clearances' took place in the Highlands. In others it took centuries to come about. The shielings have now gone, as have the flocks of wether sheep which displaced the crofters. The wether flocks have in turn been replaced by breeding (ewe) flocks or red deer; most hill areas are now stocked with breeding flocks.

Hill sheep in Britain

A large proportion of the sheep meat and wool production derives at first and/or second hand from hill and upland farms: hence, hill sheep are of considerable importance.

There is a number of different breeds of hill sheep in Britain and they fall naturally into two groups. There is a white-faced group including such breeds as the Welsh Mountain, Cheviot and Exmoor Horn. These have what is known in the trade as 'cross-bred' wool. The other group which is dark-faced and includes the Scottish Blackface, the Swaledale and the Rough Fell, have coarse fleeces of carpet wool. In addition to being coarse, the wool often shows a high incidence of kemp fibre. The primary characteristic of both these groups is their ability to live and produce offspring under adverse conditions of weather and of nutrition. They tend to inhabit areas where other domestic stock find it difficult to live, although in some hill areas hardy cattle, such as Highland or Galloway or their crosses, are farmed jointly with the sheep.

In the early days of hill sheep farming in Britain, at least in the Scottish Highlands, the production of mutton and wool from flocks of wether sheep were the main sources of income for the farmers. In the late nineteenth century, under the impact of refrigerated mutton from overseas coupled with the public's increasing preference for lamb instead of mutton, the wether flocks began to disappear from the hills. The outcome has been that all the hills now carry breeding stock including areas which, in the nineteenth century would have been considered too poor to sustain breeding ewes. The hill ewe occupies these areas by virtue of the fact that she is the only domesticated animal able to stay alive there, and this quality of live-ability takes precedence over all others. High prolificacy on the hills, however, is a handicap rather than a virtue as the nutritional status of the ewe is seldom good enough to provide milk to rear more than one lamb, and the sheep's body covering must provide protection from the elements before it can be considered as an article of trade. Mention has been made in a previous chapter of the fact that young animals are more vulnerable to diseases and more seriously affected by malnutrition than adults. Also, it is true that after a certain period an adult animal begins to decline in vigour and is more susceptible to the affects of adverse circumstances. The hill farmer is well aware of this tendency and a system — *stratification* — has been evolved to mitigate the problem.

STRATIFICATION

What the hill farmer does is to dispose of ewes after a set period, usually after three, four, or five crops of lambs. This process is known as 'casting for age'. These cast ewes are bought by farmers on lower lying grass farms where they, in turn, are crossed with longwool rams. The outcome of this mating is a female with the ideal qualities required of a meat lamb mother. The male lambs are all destined for slaughter. In some cases the cast ewes may be crossed by a small ram, such as a Southdown or even larger breeds of accept-able conformation for the direct production of small meat lambs. The Southdown X Scottish Blackface and the Southdown X Welsh Mountain produce exceptionally good small lamb carcases.

These cast ewes are not disposed of in a haphazard manner. Special sales are held in the hill areas in late summer and early autumn where the ewes are sold in tens of thousands. In addition to casting for age, the hill farmer also gets rid of any broken-mouthed ewes and ewes which, for any other reason, are not considered likely to rear another lamb on the hill. The cast ewes are used for varying periods in their new homes. How long they are kept is greatly influenced by replacement costs. Of the lamb crop on the hill all male lambs not required for breeding are normally castrated, some going ready for slaughter and the remainder for further feeding. In North Wales some farmers do not castrate as the uncastrated males can attain quite big weights without getting over-fat. Any females surplus to breeding requirements are sold to the lower ground farmers who, in turn, will tup them as gimmers with a longwool ram for the production of meat lamb mothers.

This system of crossing produces large numbers of females for use on the best quality pastures for meat lamb production. Examples of these crosses are:

- Border Leicester X Cheviot giving the Scottish Halfbred
- Border Leicester X Blackface giving the Greyface
- Border Leicester X Welsh Mountain giving the Welsh Halfbred, and
- Bluefaced Leicester X Swaledale giving the Mule.

These cross-bred ewes are further crossed with Down rams to produce meat lambs or store lambs for feeding as hoggets. In the case of the Down crosses it is normal to send both male and female lambs for slaughter but in some cases such as the Suffolk X Scottish Halfbred the females may be retained for use as meat lamb mothers. They are particularly useful where winter housing is practised as they are very docile.

It will be seen that much of the meat lamb production of a country like Britain is based on the sheep flocks of the hills. The inter-relationship between these basic flocks and the final product of meat lamb from the lowland farm is known as *stratification*, meat lamb being the focal point of British sheep production. This phenomenon of stratification is represented diagrammatically in Fig. 4.1.

Stratification confers a number of advantages on the sheep industry: the best use is made of land, the different breeds of sheep and their various strengths. The mountain ewes are thrifty, hardy and, under conditions of good nutrition, good milkers. The longwool breeds are prolific, good milkers and they increase the body size of their progeny. The final crossing sheep, the Down breeds, give good conformation, quick fleshing and quality meat. Also, the offspring of the crosses normally exhibit hybrid vigour (positive heterosis) and thrive well from birth to slaughter. 'Hybrid vigour', or positive heterosis is the term used for the fact that interbreeding often results in offspring having better performance than either parent. Stratification also enables breeders to concentrate their attention on a limited number of objectives and to use their land for the farming system to which it is best suited. Generally speaking, it is a waste of high quality land to use it for stock rearing.

It will, however, be obvious to anyone who is familiar with the

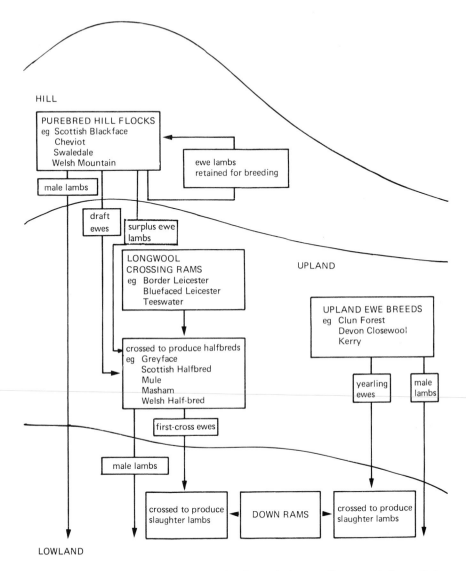

Fig. 4.1 Stratification in the British sheep industry (*by permission of the Meat and Livestock Commission*).

sheep industry in Britain that stratification is not necessarily clear-cut and distinct. In the Scottish Borders it is not uncommon to find large farms where pure Blackfaces are run on the high ground, the cast ewes being crossed with the Border Leicester lower down, and finally the resulting Greyface ewes are run on the rotational land in the valley bottom producing meat lambs sired by one of the Down breeds of rams.

In some areas of the North of England one finds rough grazing on the lower hills carrying Swaledale ewes — half of the flock being put to the Swaledale ram for breeding replacements while the other half is put to a Bluefaced Leicester or Teeswater. A specialised section of the stratification system is the production of rams. These purebred flocks are, in most cases small, some of them very small. The only purebred ram-producing flocks which are large by British standards are the Suffolks. Most of the longwool purebred flocks are small, being from about 12 to 70 ewes in number, but the people who own these flocks are often active in commercial sheep farming in addition to their pedigree work.

Two final points must be made concerning stratification. First, it is a system which was not consciously constructed by men. It evolved over a period of time due to economic and other pressures acting on the farmer and his flocks. Secondly, stratification is not a phenomenon which is peculiar to Britain. It occurs in most other sheep-producing countries about which more will be said later.

It would be wrong to suggest that all meat lamb production in Britain is based on stratification. There are grassland breeds of sheep which are self-replacing and have high prolificacy, good milk yield and also produce very acceptable meat lambs. Typical breeds from this category are the Clun Forest and Kerry Hill, and these are numerically quite strong.

The Spanish Merino

Spain was mentioned a little earlier in the chapter and it is to Spain we must momentarily return. No doubt like other European countries, Spain had a number of varieties of sheep, but from the fifteenth century a special breed, known as the *Merino*, was developed. These sheep produced fleeces of compact, very fine wool the quality of which far surpassed that of all other breeds.

These Merino flocks were very large and they belonged to the élite of Spain, the owner of one of the largest and most famous flocks being the King. In addition to the nobility the Church was also a large-scale owner. The monopoly Spain had in this outstanding wool was closely guarded and the death penalty was imposed on anyone smuggling Merino sheep out of the country. This held until the second half of the eighteenth century when the King allowed sheep to go to Saxony. Later, Merinos went to France and ultimately to Britain and various other European countries. In Saxony an exceptionally fine-woolled sheep was developed which, in due course, was to exert its influence on the Australian Merino. The French importation was developed over the years into a new breed known as the Rambouillet, which also played a part in the foundation of the sheep stocks of Australia, the Americas and South Africa.

The story of how the Merino and its derivatives were used as the foundation of the major sheep population of the world is long and complicated, but it is well documented and worthy of study, not only by agriculturists but anyone interested in economic history.

In the late eighteenth and early nineteenth centuries King George III imported substantial numbers of Merinos into Britain and these spread even as far as Sutherland. Although some good flocks were formed in the country they failed to find general acceptance and finally petered out. Various reasons have been offered for their failure to establish but, as is common in such cases, the breed failure was probably caused by a number of factors. The author, having eaten Merino-derived mutton in war-time South Africa and having been responsible for feeding cross-bred Merino wethers in Scotland, can suggest two good reasons for their failure. As a mutton carcase the Merino and its crosses leave a lot to be desired, and the animal is very prone to foot rot in high rainfall areas, especially if the drainage is poor. It is interesting to consider the distribution of the breed: although it is the most imported foundation breed of the world's commercial sheep it is not well-established in areas with a maritime climate. It is a sheep for continental conditions, being well able to withstand the extremes of heat and cold and shortages of food and water, which it does better than most European breeds. It is not now found in the British Isles, Holland or the maritime provinces of Canada, and while it was the first breed of sheep introduced into New Zealand, it is now confined to the high country of the South Island. In New Zealand their place was largely taken by Romney Marsh sheep which the New Zealanders have transmuted into a new breed.

It could well be thought that in nineteenth century Britain with its expanding demand for high quality wool to service its growing textile industry a place could have been found for Merino sheep in spite of the breed's limitations. The trouble was that the breed had already been introduced into South Africa, Australia and South America — countries which were well-suited to sheep farming and wool production.

Wool is one of the most valuable commodities per unit weight produced by agriculture. It is easy to handle, and provided it is kept dry, does not readily deteriorate. In consequence, transportation by ox wagon and sailing ship over great distances was quite possible and could be very rewarding economically. Settlers opening up new countries, such as Australia, were thus encouraged in their sheep farming and they undercut the British wool producer. Meat, however, was a very different case. Commercial refrigeration had not yet arrived; prosperity was increasing in Britain with a greater number of the artisan class having the purchasing power to satisfy an appetite for meat. The response of the farmer was not unnaturally to favour meat production as against that of wool. The Merino might have been relegated to the high hills, as it was in New Zealand, but the instinct for close flocking, so characteristic of the breed, was not appropriate to British hills. Such predators as wolves had vanished from these hills long before man populated them with sheep, and the flocking instinct and the ability to walk long distances for water and pasture which makes them easy to shepherd and protect in the wilderness were disadvantages in our hills. The grazing of mountain areas such as the Scottish Highlands is much better carried out by a breed which tends to scatter as does the Blackface.

World trade

The 'New Countries' continued to increase their exports of wool, hides and tallow to the Old World until the late nineteenth century when refrigeration was introduced. This brought about large-scale changes in sheep production especially in New Zealand and Great Britain and to a lesser extent in Australia. The Dominions responded by exporting first frozen, and then chilled carcases. These carcases came from longwool X Merino crosses with the addition of Down breeds for a final cross, another typical example of stratification. New breeds also came into existence based on longwool X Merino crosses. One of the best known of these is the Corriedale, based on crossing Lincoln and Leicester rams onto the basic Merino ewes. The New Zealand farmer, especially in North Island, turned to the Romney Marsh as foundation stock, the ram used for the production of the export meat lamb being the Southdown. This export trade from the Antipodes to Britain has endured for almost a century, but the advent of the Common Market in Europe will create yet another change, with New Zealand looking to the Middle East,

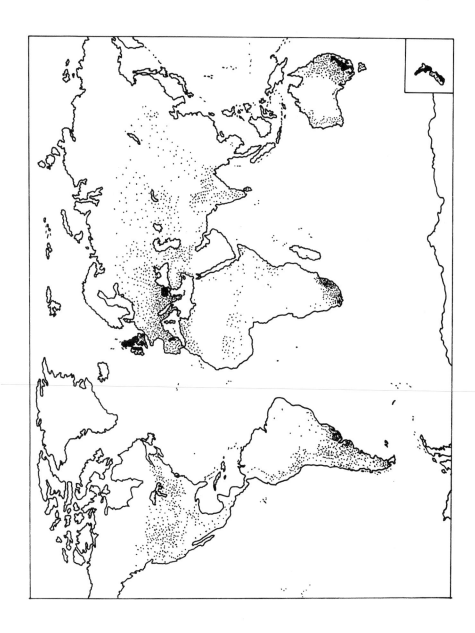

Fig. 4.2 The world's sheep distribution. (*After Fraser and Stamp.*)
Each dot represents 100 000 sheep.

America and Japan for markets.

While countries of the Southern Hemisphere have come to dominate world trade in wool and mutton, other countries have conducted their sheep husbandry on a much smaller scale, often of peasant-style farming, and in South East Europe and the Near East the sheep is still a triple-purpose animal. There are also some very specialised forms of sheep production in Asia and Africa, of which the most famous is the production of Astrakhan fur. This fur comes from the skin of Karakul lambs which are slaughtered shortly after birth.

Finally, it should be mentioned that Western Europe as a whole, apart from Spain and the British Isles, is not very interested in sheep. They do not keep great numbers and little lamb and mutton is eaten. North America is another area which has shown no great enthusiasm for sheep, beef and pork being preferred to mutton, but this situation could well change with a better effort at marketing an appropriate carcase. The main sheep concentrations are in the Range states of the west and here the foundation stock has been Rambouillet and other Merino derivatives. Russia has a large number of sheep, their main interest being wool production, but their sheep farming makes little impact on world trade. The main interest they provide to outside sheepmen is their extensive use of artificial insemination in flock improvement.

The distribution of sheep throughout the world is shown in Fig. 4.2.

Having presented this outline of the general sheep situation it is now time to turn to the actual mechanics of sheep production. These will be discussed in the following chapter.

Points to remember

1. Basic breeds:

    The *Merino* provided a very hardy breed with superlative wool, especially suited to semi-arid conditions in the 'new' countries of Australia, the Americas and South Africa.
    In Britain, hill breeds such as *Scottish Blackface* and *Welsh Mountain* provide small hardy animals able to flourish in extremely adverse conditions. Longwool breeds such as the *Leicester* and *Lincoln* provide crossing tups for hill breeds, conferring increased size, greater milking capacity and sometimes prolificacy on their offspring. Down breeds such as *Southdown* and

the *Suffolk* can be crossed on to females of the above crosses to produce progeny with improved carcase quality and growth rate.

2. Stratification:

   This systematic cross breeding on a regional scale is called *stratification*. Basic breeds of hill and semi-arid areas are crossed with a longwool ram and the female progeny are used as meat lamb mothers when crossed with Down or similar meat sires.

3. Advantages of stratification:

(a) Land is used for the most appropriate type of production.
(b) Each breed or type of sheep is used to its best advantage.
(c) Advantage can be taken of hybrid vigour (positive heterosis).
(d) Farmers can specialise and hence increase efficiency.

5 The shepherd's calendar or shepherd's year

There is a certain minimum number of operations which have to be carried out in any breeding sheep flock. These are cyclical in nature and in Britain, as in many other countries, their timing is more or less imposed by climatic conditions.

Making up the flock (late summer—early autumn)

The most appropriate place at which to start the calendar is at the making up of the ewe flock ready for the breeding season. Under normal conditions of spring lambing this is done in late summer. It is assumed that obvious rejects such as barren ewes and ewes that have had mastitis will have been eliminated earlier in the year. The flock is now examined for 'broken mouths', signs of mastitis not previously detected, serious lameness or physical condition well below the normal for the flock. These animals are rejected and disposed of. If the flock is maintained on a regular age basis the over-age ewes are drafted while any poor milkers (which are identified by having poorly grown lambs) should also go. Gimmers enter the flock at this time. They, too, should be closely examined and checked for physical disabilities such as overshot or undershot jaws and they should be of a size, weight and general conformation appropriate to the breed or cross to which they belong. In flocks which are recorded, identification marks such as ear tags should be checked and any missing ones replaced.

In order to start off the flock literally on the right foot this is also an opportune time to ensure that foot rot is well under control.

FOOT ROT CONTROL

All sheep should be turned up and each foot carefully inspected. Using a pair of high quality steel secateurs and a sheep foot knife all overgrown horn should be removed from each affected foot.

Paring should be carried out until all suppurating tissue has been exposed but cutting down into the live tissue should be avoided. The treatment of individual feet with the antibiotic chloramphenicol applied by aerosol can be very effective though expensive, but needs a veterinary prescription in the UK.

The sheep whose feet have been subjected to surgical treatment should be removed from the rest of the flock and stood on a clean surface, such as concrete, to dry off. The sound-footed sheep should be put through a 5% formalin footbath and placed in a clean field which has not carried sheep for a fortnight. The sheep under treatment should be held on a concrete yard for two or three weeks and be subjected to the formalin footbath twice a week before returning to the main flock. In areas where foot rot is a serious problem, the flockmaster should consult his veterinary surgeon with a view to the use of foot rot vaccine.

Tupping

Whether or not the ewes are flushed will depend on the system of farming and the condition of the ewes. The heavy flushing of hill ewes is not desirable as multiple births are not wanted on the hills. Well-fleshed meat lamb producing mothers do not need intensive flushing but they need to be on sufficiently good keep for them to rise in weight throughout the tupping period. On no account should a ewe be allowed to fall in weight during and immediately following tupping. The most satisfactory medium for flushing ewes is a reseeded pasture.

Repeated reference has been made in the book to the *condition* of sheep, animals being spoken of as being in poor, moderate or good condition. These statements, although very subjective in character, convey quite a lot of information to persons knowledgeable in sheep husbandry. A successful attempt, however, has been made in Australia to quantify information on condition, based on a scoring system. It has been taken up by research workers in Britain and its use is being encouraged by the Meat and Livestock Commission. Briefly, the scheme is as follows:

The score ranges from 0–5 in an ascending order of fatness. Scoring is done by examining the lumbar region of the sheep by touch:

- *Score 0* This represents extreme emaciation – literally skin and bone.

- *Score 1* The spinous processes are sharp to the touch as are the transverse ones with very little eye muscle present.
- *Score 2* The spinous processes are fairly prominent but not sharp and the transverse processes are smooth and round. The eye muscle is moderate with a little fat cover.
- *Score 3* The spinous processes show only as lumps, being smooth and rounded. The individual bones can be felt when pressure is applied. The transverse ones are well covered, the eye muscle being full with moderate fat cover.
- *Score 4* The spinous processes can just be detected with pressure. The ends of the transverse processes cannot be felt. The eye muscle is heavily covered with fat.
- *Score 5* This is a condition which should never be found in any sort of modern sheep. The spinous processes cannot be detected with heavy pressure, nor can the transverse ones. The eye muscle and the tail of the animal is heavily over-laid with fat.

Two factors will be readily apparent from this table. The first is that in many descriptions half points will, of necessity, be used. The second is that the animals described at each end of the list are quite unsuitable for breeding. The ewes should not score less than 2 or more than 4 when put to the ram. For hill ewes the score should be 2½–3; for low ground ewes the score should be 3½–4.

Inoculations

The period of setting up the flock for tupping is also a suitable time for vaccinating the ewes against clostridial diseases. This is best done by giving a combined vaccine which covers such diseases as lamb dysentery, pulpy kidney, tetanus and similar diseases. The manufacturers of vaccines issue advice on the actual practice and timing of injecting sheep, but it will not come amiss to underline them here. As the injections are subcutaneous, the first requirement is a suitable number of short, sharp needles. These, together with the syringes, must be sterile. Boiling in water for a quarter of an hour will achieve this. The best site for the injection is the side of the neck so as to avoid damage to the carcase (see Fig. 5.1).

The sheep should be dry when they are injected otherwise contamination of the injection wounds may occur. The routine is quite simple:

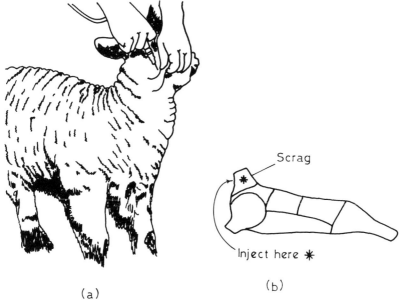

Scrag

Inject here ✳

(a)   (b)

Fig. 5.1 To avoid carcase damage, inject in a low-priced area such as the neck. Most flock masters never see the carcase mutilation which is necessary to remove vaccination abscesses from high-priced areas (*Meat and Livestock Commission*).

- Check the label and maker's instructions to ensure that the correct vaccine is being used and in the right dosage.
- Wipe the sealing cap of the vaccine bottle with cotton wool soaked in surgical spirit.
- Take a syringe complete with needle and draw into it a little air, push the needle through the bottle cap and expel the air to prevent the formation of a vacuum.
- Withdraw a little more than a complete dose of vaccine, disengage the syringe and fit another needle to it.
- Adjust the dose to the correct volume and see that there are no trapped air bubbles in the syringe, take a fold of the skin on the neck and inject the dose.
- The needle in the bottle should remain there until the bottle is empty. This prevents contamination of the vaccine.

Multi-dose disposable equipment is now becoming popular and, when used according to makers' instructions, reduces the risk of infection. Frequent changes of needle are necessary, about every 25 sheep.

At the end of the day any remaining bottles which have been

opened, but not fully used, should be destroyed. On farms where ewes are kept intensively this gathering of the sheep also provides a useful opportunity for worming. Finally, before introducing the rams, the ewes should be clipped round the base of the tail. This can be of importance with heavily fleeced ewes.

Many farmers find it helpful to run one or two vasectomised rams with the ewes for three weeks before the fertile tups are introduced. These rams seem to stimulate the ewes to come into season thus shortening the period over which the ewes are got in lamb, as a larger proportion of the ewes will therefore be in season than is normally the case when the fully functional rams are turned in.

The ram

Rams should be checked in a similar manner to the ewes, particular attention being paid to the hind feet as a ram which is lame at tupping time is most unlikely to perform in a satisfactory manner. The ram should be bright-eyed and alert with a well-poised head and a somewhat aggressive mien. Rams should come from a breeder of repute and, where possible, he should be local. The breeder's reputation should also be supported by test figures for his rams wherever this is practicable. This is particularly important in small flocks where the failure of a tup to perform adequately needs to be remedied at short notice by the prompt introduction of a replacement. The ram should be true to type, well-grown for his age, free from blemishes and have two uniform, properly descended testes. In pure-bred white-woolled flocks the presence of black wool at the neck or dock should be avoided, as this could lead to a down-grading of fleeces.

The time at which tupping starts will depend on factors which will be discussed under 'systems' in Chapter 9 as will the number of ewes allocated per ram. The ram should start the breeding season in a good, well-fleshed but not fat, condition so that he spends his time seeking out receptive ewes and not in looking for food. The ram should nonetheless receive some hand-feeding while working.

The number of ewes allocated per ram varies: while a ram lamb or enclosed ram will be given no more than 30 ewes, a fully mature ram will have 80 or more. Under range conditions fewer ewes must be allocated per ram.

COLOUR MARKING OF EWES AT TUPPING

In low ground flocks it was once the practice to mark the briskets of the rams with a colouring material such as yellow ochre and linseed oil. This colour was transferred to the rumps of the ewes at service, thus showing the extent of sexual activity in the flock. This method of colour marking has now been largely superseded by the use of a crayon which fits into a metal holder attached to a 'mating harness' carried by the ram. The crayon colours are changed every 16 or 17 days and, by this method, it can be seen whether or not tups are working as, if they are not, a large number of ewes will repeat. When such a case arises the ram or rams are replaced. It is sound practice to change the rams during the breeding season and with large flocks, rams can be kept in reserve to go into the flock at the first colour change. It is not desirable to have only one ram with a group of ewes because, as with other animals, a ram can have his 'off days'.

Colour marking the rams has a further advantage in that it enables the shepherd to divide up the flock into suitable groups for supplementary feeding and lambing. The importance of the supplementary feeding of the ewes being closely related to the lambing date often receives too little consideration by many farmers. As a final note on *raddling* (as the colouring of rams is termed), the easiest sequence of colours to read is yellow, blue and red, green and purple.

This is an excellent point at which to reiterate that sheep are run on a *flock* basis and not on the basis of the individual animal. It is to the flockmaster's advantage to concentrate the lambing into as few days as possible and, to this end, every effort must be made to see that all ewes are served by a fertile ram at the first opportunity.

Conception and pregnancy

When the ewe has been served and the eggs fertilised, the eggs do not become immediately attached to the uterus. They have to be sustained from nutrients in their fluid surroundings. Attachment to the uterine wall takes place in the third week after service when the placenta starts to develop and the embryo begins to satisfy its nutritional requirements from the bloodstream of the mother.

A significant embryonic loss in the ewe is thought to take place during the first month of pregnancy, but this is difficult to demonstrate and quantify. The early stages of placental development are critical and a good supply of nutrients is essential if development is not to be impaired: any damage done by deprivation at this stage cannot be put right by better feeding later. As well as lambs being lost early in conception, impaired nutrition can also lead to uneven development in twins with one being born weakly.

It follows, therefore, that anything that is likely to have an unfavourable effect on the ewes should be avoided:

- Ewes must be kept on a good plane of nutrition until some time after tupping finishes. Tupping should, therefore, take place on a good pasture where the ewes will not go back in condition.
- The field, or fields, should be of adequate size and secluded where the sheep are free from harassment by dogs, people or traffic.
- The ewes should not be overcrowded or in any way brought under stress.

If these precepts are kept a good conception rate should be achieved.

The normal time to leave the rams with the ewes is 42 days or so but in some small flocks the rams are allowed to run on with the ewes. In larger enterprises where operations are definitely on a flock basis the former time schedule should be adhered to as it gives the ewes at least two opportunities of being served. As the average oestrus cycle is about 16 days the majority of ewes get three opportunities of receiving the ram. Leaving the rams in longer will give rise to a small number of lambs which will be out of phase with the rest of the lamb crop. Their mothers will also be out of phase with the rest of the flock and this could be the start of the disruption of an orderly production system.

For the first three months of the 5-month pregnancy period the embryos do not make a great demand on the ewe, and during the second and third months all that is required is to keep the sheep well-fleshed and not let them go back in condition. They should not, however, be allowed to get fat. Feeding the ewes up to, and after, lambing will be dealt with under 'systems' in Chapter 9.

A problem that can arise with heavily fleeced ewes in later pregnancy is that they get on to their backs and are unable to recover their feet. This is particularly common on uneven land. Old, heavy land pastures in ridge and farrow were particularly prone to such accidents. Any ruminant animal which gets on to its back and is unable to recover usually becomes bloated in a short time and, if not attended to, dies.

Lambing

Most grassland flocks are lambed in the spring. In Britain, February, March and April are the main months, depending on climate and husbandry conditions. In outdoor lambing, the greatest danger is from adverse weather although under some range conditions predators, such as wolves, dogs and foxes can wreak a great deal of havoc. The worst weather young lambs can face is wet cold. Once the lamb is dry and has had a good feed of colostrum, the battle for survival is well on the way to being won.

PREPARATION FOR LAMBING

In view of the weather risk every effort should be made to take advantage of any type of shelter available, both natural and artificial. The form of protection given can vary widely — a shelter belt of spruce trees, a dry stone wall, a pen of straw bales, or the ewes and lambs can be fully housed. Small flocks kept in small fields present no big problem but with large flocks of low ground sheep or big ranges, enclosures are necessary. In the past it was the practice on many farms to have permanent lambing yards. These were well adapted to easy handling of the ewes and lambs but tended to harbour diseases such as joint ill. A more satisfactory method of dealing with the situation is to build temporary pens with poles, wire netting and straw bales. These pens can be used for a season or two and then burned down. The site for a temporary set of lambing pens should be on well-drained, rising ground which, in the northern hemisphere, should be sheltered from the north and east. It needs a water supply, and electricity is highly desirable though not essential. Alternatively, a building in the farm steading can be used for lambing and here the major requirement is that it is easily cleaned and disinfected. In either case, small pens where ewes in difficulty and ewes which are having lambs fostered on to them can be confined are a necessity. A useful size of pen is 1.5 m X 1.3 m with sides 1 m high.

In some flocks where the sheep carry a heavy and all-embracing fleece, udder locking is practised before lambing. This is the removal of wool from around the udder so that lambs may find the teats without difficulty and without having their mouths filled with wool. Whether or not udder locking is practised as a special operation on the flock, each ewe should have the wool cleared away from the udder at some time. In most flocks it is sufficient for the shepherd to pluck away any excess wool when he checks the milk supply of the ewe and her freedom from mastitis immediately after lambing.

Final preparations for lambing are completed by the shepherd acquiring a supply of buckets, clean towels and ample soap. Buckets are required for hot water needed at assisted lambings and also for watering individually penned ewes. The importance of providing ample water for ewes immediately after lambing cannot be exaggerated as deprivation for any length of time will lead to a marked reduction in milk yield, and could also lead to the weakening of the maternal bond with a consequent ill effect on the lamb or lambs.

The lambing process

If the flock is of any size it will need to be divided into at least three sections in accordance with the colour marking of the rams. This is in order that the feeding should be appropriate, and also that the ewes come into the lambing quarters in the correct sequence. Ewes should be moved into their lambing quarters a few days before the first ewe is due to lamb so that they can settle down and familiarise themselves with their new surroundings.

When the first ewes are due, a careful watch should be kept in order to check how long any ewe has been in labour. Labour can take anything from a matter of minutes up to four hours before a normal birth occurs and, unless the membranes have ruptured, the ewe should be allowed some hours before assistance is offered. In normal circumstances the ewe lambs herself, licks the lamb dry and it is soon on its feet and suckling without outside assistance. The usual indication of help being needed is the appearance of a head without the legs, or only one leg showing, or a breech presentation. Indeed, the number of malpresentations possible is quite large.

If it is decided that the ewe needs assistance the first thing that the operator must do is to wash his hands and arms thoroughly in hot water using ample soap. Ensure that there is a good lather as this is an excellent lubricant. The ewe's vulva must be washed and then a careful exploration of the foetus or foetuses be made and of their position in the uterus. Fig. 5.2 shows two typical examples of how lambs may be positioned in the uterus. With a little knowledge of anatomy and a little thought it is not difficult to work out by feel the position and posture of the lamb or lambs. With a little patience it is normally possible to rearrange the unborn lambs so that they can be safely delivered without too much damage to the ewe. The

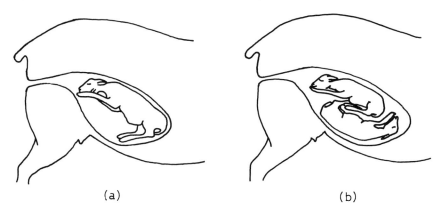

(a)   (b)

Fig. 5.2 (a) Normal presentation of the new-born lamb. *(b)* Twin pregnancy showing both normal and breech presentations. (*After D'Arcy.*)

lamber must always bear in mind that the uterine wall is quite delicate and a carelessly moved lamb's foot can be the cause of a fatal rupture of the uterus.

There are, however, a number of conditions and abnormalities which will defeat even the most competent shepherd and a decision has to be made as to whether a veterinarian should be called or the animal slaughtered. One such case is *ring womb* where the neck of the womb does not dilate sufficiently to allow the lamb to be expelled. The decision to call the vet should be made promptly as it is most unfair to call for professional assistance after the animal has been exhausted by unskilful and misdirected pushing and pulling.

It is important that once parturition has started it should be completed as quickly as possible as the more protracted the lambing the weaker the ewe-lamb bond tends to be and the sooner the ewe and her lamb are penned together the faster and more firmly will the bond be established. Where the ewes are in a field there should be ample room for the ewe to find a place to mother her family undisturbed. The avoidance of disturbance by other ewes or extraneous forces should be one of the shepherd's first priorities.

Protracted lambing not only weakens bonding but also has an adverse effect on the physical condition of the ewe which, in turn, will probably have an adverse effect on milk production. There is also the point that, in the case of a breech presentation, the lamb may start breathing before the head is disengaged and fluid may be inhaled into the lungs.

The crucial role of the shepherd during the act of giving birth is to ensure that the newly-born animal is breathing. When a lamb has been delivered the muzzle should be cleared of membrane and mucus. If it does not start breathing within a few seconds, the operator should blow down the lamb's throat in order that the carbon dioxide in his breath can trigger off the breathing mechanism of the lamb. Artificial respiration, performed by placing the lamb on its side and simulating the breathing action in the lamb by raising and lowering the fore limbs, often gives the desired result.

Lambing will not be discussed further as it is an essentially practical craft to be learned under expert guidance in the lambing field rather than from books. In Britain students can get very good preparatory instruction at Agricultural Training Board Centres where instructors use artificial uteruses to show various malpresentations and the proper method of resolving them. In most countries there are agricultural colleges, young farmers' organisations and similar bodies which run courses on such practical matters as lambing and all students should be encouraged to take advantage of these facilities.

MIS-MOTHERING

The problem of mis-mothering can be quite serious and can arise from a number of causes. In well-tended lowland flocks it is usually caused by unlambed ewes with a highly developed maternal instinct poaching a lamb from a ewe lambing triplets, or even taking an only lamb from a gimmer. It is for this reason that lambing flocks need a lot of room or a lot of attention. The second cause is malnutrition of the ewe. Here the thin ewe with twins and a low milk supply tends to go off with the stronger lamb and leave the weaker. Finally, there is the case where a lambing flock is attacked by a predator or predators, such as dogs, wolves, coyotes or the like. The flock is panicked by the attack, the lambs get mixed up and separated from their mothers and die before they can find their own mother and get fed.

FOSTERING

It is desirable that an attempt be made, where possible, to break up sets of triplets and foster one of the lambs on to a ewe with no lambs or one with a single. In flocks with a high lambing percentage it may not be very easy to find foster mothers for broken up sets of triplets or quadruplets, as the majority of ewes will have at least two of their own lambs to look after. Many prolific, high-milking ewes such as mules and greyfaces will rear triplets quite readily but these ewes need close attention and very good feeding.

In fostering, always take the strongest lamb as it is the one best able to come to terms with a foster mother which may not be too amenable to its attentions. Also, the lamb should be placed with the recipient ewe as soon as possible. Smearing the lamb with the ewe's uterine fluid is quite often all that is needed. The bond between ewe and lamb appears to be based chiefly on smell and sound. Where the recipient has a dead lamb, the lamb can be skinned and its pelt secured to the fostered lamb in the manner of a waistcoat. This is a very satisfactory method of ensuring adoption.

Fig. 5.3 Proprietary lamb adopter (*available from R L Farmer Ltd, Doncaster, UK*).

It is now possible to obtain proprietary lamb adopters built of wood. These have provision for four ewes to be yoked facing inwards towards each other in a cruciform manner (see Fig. 5.3). There is a space for the lambs on either side of each ewe and the ewe is so held that she cannot attack the lamb or lambs. The ewes and lambs should be penned together for a few days until the lambs are fully accepted. In addition to having individual pens for ewes and lambs in trouble, where electricity is available it is highly desirable to have infra-red lamps. When subjected to the influence of the heat lamp the majority of lambs that have become wet and chilled stage a prompt recovery.

Colostrum

Colostrum, or beastings, is the first fluid produced in the udder of the ewe on parturition. It differs from normal milk in a number of ways. It is a thick viscous liquid, high in vitamins and protein. Associated with the protein are antibodies against diseases with which the ewe has been in contact, or against which she has been inoculated. It is also laxative in character, and serves to rid the lamb's intestines of foetal meconium. In other words, it has the effect of flushing out the gut ready for the lamb's digestive system to go into action. A good intake of colostrum provides the lamb with a high energy source to provide heat and do work, a high and readily digestible protein source with which to build up its tissue and also the means to defend itself against disease. It follows from this that the shepherd's first job, once the lamb is breathing, is to see that it does receive a full feed of colostrum. This is not only because of the importance of an early start for the lamb, but because the ability of the lamb to absorb antibodies falls off sharply after the first feed, unlike the calf where this ability declines more gradually with time. For this reason, the shepherd should ensure that the first feed of colostrum a lamb gets is adequate. If the mother has insufficient, the lamb should be 'topped up' from a reserve supply.

Colostrum may be milked from ewes as opportunity offers and stored in a refrigerator until it is needed. It can then be administered by means of a stomach tube, taking care not to insert the tube in the lungs. These tubes can be bought complete with a glass reservoir at any veterinary product stockist.

This question of adequate colostrum intake brings one back to birth weights and it is evident that the smallest lamb in a set of triplets will be at a marked disadvantage compared with its stronger sibs, and it is here that good shepherding makes a major contribution to the successful rearing of a large lamb crop. In view of the fact that the majority of deaths amongst lambs are perinatal and that only about a quarter occur after lambs are a week old, constant reiteration of the importance of care at lambing time needs no apology. With well-shepherded lowland flocks the death rate may be as low as 5% or even lower, but a more general figure would be

8–10%. On high hills under severe weather conditions the losses can go well above 30%.

After-lambing care

As soon as the lambs are strong on their feet and the bond well-established between themselves and their mothers, they should be removed from the pens or lambing fields and moved on to pasture fields. In situations where there is a large number of big litters which are not being broken up, it is useful to mark the ewes and lambs with a marking stick so that any mixing or mismothering can be recognised at a glance. The majority of ewes are not tolerant of cross-suckling and in a congested area it is easy for a lamb to vacillate between two or more ewes and finally become rejected by all. If the ewes are going on to pasture and grass is in short supply, supplementary feeding will be necessary. Ewes rearing twins or triplets should receive about 450–650 g (1–1½ lb) concentrates per day. A few mangels, swedes or potatoes are a great help in maintaining milk production. These vegetables seem to have a stimulatory effect far beyond their nutritional content. It is also important to have an adequate water supply available to the ewes. A close watch for signs of ill health must be kept on both ewes and lambs. For the first week after turnout a progression of nursery paddocks is ideal since many lambs are lost immediately after turnout due to neglect.

Mastitis is one of the commoner problems at this time and any ewe which refuses to let a lamb suckle should be turned up and examined for udder trouble. If any suspicious deaths occur in either ewes or lambs veterinary advice should be sought promptly. Where creep grazing is not practised the flock should, where practicable, be divided up on the basis of the best feed to the greatest need. Ewes with twins or triplets should be put on the best pastures, gimmers with singles on the next best and older ewes with singles on the poorest. In areas where *Nematodirus* is present, lambs should never be run on other than maiden leys. In spite of every effort by the flockmaster and shepherd there are bound to be some losses due both to still births and early deaths. Whilst it is important to improve husbandry and techniques so that these are minimised it is well to consider what can be salvaged from the inevitable losses. In some areas of New Zealand this matter has been taken in hand and slink lambs are collected for the factory processing of pelts. About

1 500 000 skins from this source were processed there in 1977. This is not to suggest that New Zealand has higher losses than other sheep keepers but merely indicates that they are prepared to do something about the losses, and others would do well to follow.

FLY STRIKE

In many countries, especially Britain and New Zealand, considerable trouble is caused by blowflies. These insects lay their eggs on soiled wool in such places as the tail head. The eggs hatch within about three days to produce white maggots which eat their way through the skin into the flesh of the animal. These maggots feed for up to a week during which period they moult a number of times. Finally, they drop to the ground where their skins turn brown and the maggots pupate in the soil. Provided that it is still summer, they remain in the soil for about a fortnight to three weeks when a new fly emerges. In Britain the fly which causes most trouble is the *green bottle* (*Lucilia sericata*), but others are active both as primary and secondary strikers.

The green bottle is active from May onwards and the danger is most acute in close, damp weather especially in situations where there is ample ground cover such as trees, bushes and bracken. The symptoms of strike are usually obvious. The sheep stamps its feet, wags its tail violently and tries to nibble the affected parts. The blowfly is attracted to the sheep by the odour of decaying wool and other organic matter — scouring animals or those with suppurating wounds being particularly attractive.

The curative treatment is to clip the wool away from the affected part, clean the wound with a mild antiseptic and keep the wound clean until it heals; also ensure that no secondary strike takes place. Every effort should be made to prevent fly strike and, to this end, the sheep should have all soiled wool clipped away from the breech and tail (this is known as *crutching*). The animals should also be sprayed or jetted with an appropriate dip. Spraying and jetting will be dealt with in the next chapter.

In hill areas and on range ground covered with scrub or bracken, struck ewes will often hide in the undergrowth where, if undiscovered, they will ultimately die. When the fly is active it therefore behoves shepherds on such ground to ensure that their dogs search all cover closely and systematically. Fly strike is discussed at length in this chapter for the same reason as foot rot, namely that precautions against both conditions should be routine. Attacks of both are inevitable, given appropriate conditions, and preventive measures can ward off serious consequences. In all areas where blowflies are troublesome, spraying or jetting against them must be made a routine procedure.

Docking and castrating

DOCKING OR TAILING

The practice of tail docking low ground sheep is an operation which has been carried out for a very long time. It is not known when the practice started but it probably began when the folding of sheep on specially grown lush green crops came into favour. The dung of sheep fed on low fibre diets, such as brassicas or young pasture grass, does not retain its relatively hard pellet-like formation but becomes very loose and the sheep quite often scour. This semi-liquid dung soils the tail and gets clogged up in the wool. This is damaging to the wool and very attractive to the sheep blowfly.

The length of dock left depends on local fashion and on the type of sheep. In Down flocks kept for ram production the docking is very close but with ewes such as meat lamb mothers the dock is left fairly long to give protection to the vulva and udder.

In Britain the breeding stock on hill land is not normally tailed and, when it is done, only the end of the tail is removed. The methods of docking commonly employed are severing with a sharp knife, searing with a hot iron or by the application of elastic bands known as elastrators. The first method is simple and, if done within three weeks of birth, results in little bleeding. The second is normally practised on older lambs and is more or less confined to ram-breeding Down flocks. Here the lambs are older when operated on but the searing of the flesh prevents undue haemorrhage. The elastrator method has become very popular as it is simple and not as messy as the previous two. Docking with the elastrator should be carried out as early as possible: in Britain the law stipulates within one week of birth. The operation is quite simple. A set of four-pronged reverse action 'forceps' are used to expand the ring which is then placed over the tail and manipulated to the position where the tail has to be severed. The ring acts as a ligature which restricts the blood flow and in time the tail sloughs off. The first two methods of tailing expose the lambs to the danger of tetanus to a greater extent than does the third but lambs docked and castrated by the ring method are not immune from attack. Docking and castration are normally carried out simultaneously and it is a common practice to leave the tail long on *rigs*, a rig being a male with an undescended testicle. Leaving the tail long makes the lambs easy to identify at a later date, which is important when they are going to the store market.

CASTRATION

Castration is a practice which has probably been carried out since neolithic times, and is one of the earliest applications of mass selection. One deprives the inferior, or what one considers the inferior, males of the opportunity of breeding. Castration has a marked effect on the sheep as, in addition to becoming infertile and lacking in sexual drive, there is a noticeable reduction in the manifestation of secondary sexual characteristics such as growth of horns. The castrate also tends to lay down more fat than the entire male, and to grow more slowly. With the modern taste for lean and tender meat this state of affairs is not desirable: in other words, the castration of lambs intended to be sold for slaughter directly off their mothers should not be carried out. Where male lambs are not sold for slaughter off their mothers and go for further feeding, castration is necessary as their male characteristics may be so far developed by the time they reach slaughter weight as to make them unacceptable to the butcher. It may, however, be necessary to make exceptions for male triplets which take so long to grow as to need a fattening period after weaning.

As with docking, there are three common methods in use for castrating. The oldest is surgical: the bottom of the scrotum is slit, each testicle drawn down separately, the spermatic cord and the blood vessels twisted round and then rubbed through with a knife blade. In some cases of older lambs the removal of the testicle may be done with a hot iron.

The second method which achieved great popularity some years ago was the Burdizzo emasculator. Here the cords are crushed with what is, in effect, a pair of double action pincers with blunt jaws. The action of the instrument is to crush the spermatic cord without damage to the scrotum. To ensure success the testicles need to be pulled well down into the scrotum and each cord severed separately and at an angle to the line of the cords. This method is perfectly satisfactory in the hands of a competent and careful operator.

The simplest method of castration is by use of the elastrator. Here, again, it is essential that the testicles are drawn well down into the purse, the ring being placed well above them. The constriction of the ring causes both testicles and scrotum to slough off. The great advantage of the rubber rings for both tailing and castrating is speed. Under many farming systems it is at tailing and castration that the first proper count of lambs is made.

The older an animal gets, the more it is affected by pain and distress, so that all docking and castrating should be performed when the animal is very young. In Britain, any operation carried out on an animal above the age of one week requires a local anaesthetic.

Weaning

In commercial sheep husbandry it is the normal practice to remove the offspring from their dams at a given time. This is known as *weaning*. It may be when the lambs are ready for slaughter, it may be when the lambs are a specific age e.g. 12 or 14 weeks, or it may be – as is the case with hill flocks – at a specific time determined by the annual lamb sales.

Weaning is necessary in order that the ewes can be brought back into condition for tupping at the appropriate time in order that a new breeding cycle can commence. Lambs that are weaned from their mothers and go for further feeding prior to slaughter are known as *store lambs*.

Weaning is normally carried out when the lambs are twelve to fourteen weeks of age, although under intensive systems it can be much sooner – as early as three weeks. In most meat lamb producing flocks the lambs should go for slaughter straight from their mothers. These lambs are selected on the basis of their weights and how they 'touch'. ('Touch' means the palpation of the lamb at selected points to ensure a sufficient but not excessive covering of fat on the carcase.) Weaning is a simple process of removing the ewes from the lambs and out of earshot into well-fenced fields, and it should be carried out before the ewes get into too low a condition. Where store lambs are being weaned this should take place on the morning of the store sale, or as near to this time as possible, so that the lambs do not lose their bloom and still look their best. Weaned lambs tend to lose condition for a week or so after weaning and this needs to be borne in mind when buying stores. There is much in the old saying: 'you must keep them a fortnight before they are worth what you paid for them'. Weaned lambs which are retained should be kept on a clean aftermath, and it is usual to give an antihelminthic drench at this time.

A number of references have been made to drenching for internal parasites and this is an opportune place to say a few words on the subject. Drenching is normally carried out with a drenching gun with a pistol action. The drench is carried in a rubber bag or plastic container and at each pressure of the trigger a measured dose is forced through the metal nozzle which is placed down the throat of the sheep, taking care not to introduce the dose into the lungs. The operation is quite simple and effective provided a number of points are observed. The first is to check the dosage of the gun and ensure that each sheep is receiving the appropriate dose. The second is to ensure that the drench is kept agitated so that any suspension does not settle out. Finally, drenches can discolour the urine of the sheep which, in turn, discolours the wool, and for this reason drenches based on such compounds as phenothiazine should not be administered immediately prior to shearing.

Returning to the subject of weaning, the ewes should be put on to poor keep for a week or so in order to dry them off. Regular and careful checks should be made so that, should mastitis develop in any of the sheep, prompt remedial action may be taken in the form of an inter-mammary antibiotic injection.

The routine handling of the sheep flock has now been covered, with the exception of shearing and dipping. These operations together with a brief review on wool quality will be discussed in the next chapter.

Points to remember

1. The shepherd's calendar.

Flock operations	Time
Making up the flock	Late summer to autumn
Tupping	Late summer to autumn
Inoculating	Late summer to autumn & spring
Lambing	Late winter to late spring
Docking and castration	Preferably within 1 week of birth
Fly spraying and dipping	May to June start
Shearing	Late May to early August
Weaning	When lambs are ready for slaughter, or 12–14 weeks of age

2. Routine flock operations.

(a) Making up the flock:

Reject sheep with faulty udders, broken mouths or lameness.
Point score ewes and give extra food to lean sheep.
Inoculate against clostridial diseases.
Check feet and give footrot treatment.

(b) Ram selection:

Ensure tups are strong, active and free from lameness.
Colour mark and ensure that all rams are workers.
Keep some rams in reserve in case of fertility trouble.
Ensure that there is no decline in nutrition of ewes during tupping and afterwards.

(c) Lambing:

Ensure adequate supply of vaccines, medicines, soap, hot water, fostering accommodation and, as soon as possible, a reserve supply of colostrum.

(d) Docking and castration:

Perform within one week of birth if possible.
Methods: elastrator, Burdizzo, knife and hot iron.

(e) Dipping and spraying:

To prevent blow fly attack sheep must be crutched and kept clean.
Spraying and dipping should start just before the fly becomes active.

(f) Shearing:

Carry out when the wool has risen adequately i.e. the yolk is rising in the fleece and the wool has started to grow.
Shearing must be done during good weather.

(g) Weaning:

Weaned lambs for retention must be wormed and moved on to fresh clean pasture.

3. Main attributes of colostrum.

    High vitamin content.
    High protein content.
    Antibodies from mother confer passive immunity on lamb.
    Laxative effect to clear out intestine of meconium.

6 Shearing and handling of wool

Wool has been used as a material for making woven cloth since before the advent of written history but the exact when, where and how of its origins must always remain obscure.

The normal body covering of a modern sheep is far removed from the type of pelt carried by its ancestors. The original sheep probably had a coat similar to that of many other mammals – a hairy outer covering with a soft, air-retaining undercoat. The couter coat provided a thatch to repel the rain while the undercoat provided insulation. The modern fleece is the outcome of man selecting the undercoat at the expense of the overcoat over thousands of years and now fibres from the outer such as hair and kemp have disappeared from the best fleeces.

The true wool fibre of the sheep is not shed as is hair but grows continuously, although the rate of growth varies according to a number of factors such as breed and nutrition. It follows that on a sheep the wool will continue to accumulate unless it is cast by a break occurring across the fibres. These breaks are caused by conditions such as running a high temperature, suffering severe nutritional deprivation or by sulphur deficiency. Under conditions of ample nutrition and little stress the wool continues growing in a fairly uniform manner. Under harsh conditions and with poor feeding such as occur in mountainous country during winter the rate of wool growth is much reduced. Extra feeding does not elicit much response from the winter growth of wool, at least in the case of such sheep as Scottish Blackfaces.

Mention of hill breeds prompts the question of how much fleece and what type of fleece a hill sheep requires. A fleece that performs the function of a thatch and sheds the water is preferable to those which do not. While the exact type of fleece that is best for sheep kept on the high hills of Britain may be debatable, coats which trail the ground and get balled up with snow or that are parted down the spine allowing the rain direct access to the skin are obviously unsuitable for these conditions.

While the protective aspect of the fleece takes pride of place in mountain sheep the economic aspects of wool must be borne in mind especially in low ground sheep. The question has to be asked whether the wool is meant to be sold as a primary product as in the case of wether Merino flocks in Australia or as an important secondary source of income as in the case of hill farms or, as in some small lamb producing flocks, whether it is merely a minor nuisance to be got rid of as easily as possible. If the sale of wool is an important factor then careful notice must be taken of market requirements and, to this end, we will take a look at wool as a raw material of the textile trade.

Wool quality

The quality of wool is dependent on a number of factors which will be described briefly.

FINENESS

One of the most important qualities is fineness or narrowness of fibre diameter because fine fibres are necessary for spinning fine yarns for lightweight cloth. The universal standard of fineness was, until recently, the Bradford Count or Quality Count Number which is defined as the number of hanks of yarn, each 560 yards long, which can be spun from one pound of top or combed wool. This standard is still used but the major wool countries now express fineness in terms of the thickness of the fibre in microns (1 000 000 microns = 1 metre). As previously stated, the Merino has by far the finest wool and the count for the breed is from about 64 (21 microns) and upwards. The finest woolled of the Down breeds, the Southdown, has a count of about 56–60 (27–28 microns) but with wools of carpet type, such as the Scottish Blackface, the count would be 40 or below (38 microns or more).

Coarse wool fibres grow from large primary follicles in the skin (see Fig. 6.1). The fineness of wool, however, depends on the number of secondary follicles which are also present: these are similar to primary follicles but are smaller and produce finer fibres. The number of such follicles varies enormously between coarse-woolled hill sheep and Merinos, the Merinos having ten times the number of secondary follicles per unit area compared with a Scottish Blackface. Even a longwool such as a Leicester has far fewer secondary

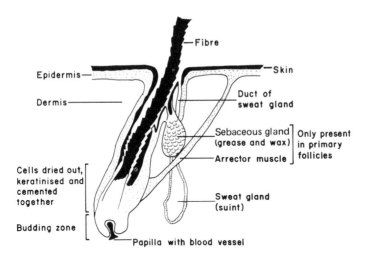

Fig. 6.1 Cross-section of a wool follicle.

follicles than a Merino. The *ratio* of secondary to primary follicles determines the type of fleece produced by the sheep.

One practical point must be noted with regard to fineness. Fleeces are not uniform over the whole sheep, being finer at the shoulders and back than at the breech. For example, a long woolled sheep would have a shoulder count of about 44 (35 microns) with a correspondingly greater breech. This variation in uniformity is much less marked in specialised wool breeds such as the Merino than in such animals as hill sheep. Also, the wool of rams tends to be coarser than for other sheep. Finally, sheep kept under conditions of limited nutrition as in some semi-arid areas tend to produce finer wool than similar sheep kept on lush pastures.

LENGTH

The next most important quality required of wool is a uniform and suitable length. This is particularly important in the manufacture of worsteds in which process the fibres are laid parallel to each other before they are spun, thus giving a strong hard thread. Uniformity of length is important because short bits of fibre need to be combed out before the wool is spun.

STRENGTH; ELASTICITY; CRIMP

Other considerations include strength and elasticity together with crimp or waviness in the staple. Sheep like the Merino show a marked crimp which gives good elasticity but strength is mainly a matter of absence of breaks or weak points caused by ill health, starvation, sulphur deficiency or other stress factors. These breaks in the wool give rise to short bits which have to be combed out. The same is true for double cuts, caused by faulty shearing, which are amongst the commonest defects in fleeces.

COLOUR AND BEHAVIOUR WITH DYES

The final points of importance when considering fibre quality are colour and behaviour to dyes. It is obvious that the darker the colour of the wool the greater the restriction on the range of dyes which can be applied successfully to it. It is for this reason that black or red fibres are so objectionable. In addition to the colour problem there is the problem of kemp and other medullated fibres which will not take dye and while a number of such fibres are normal in a tweed suiting they are unacceptable in the majority of cloths. Fig. 6.2 illustrates the different types of fibre present in a fleece. The photomicrographs in Fig. 6.3 give some idea of relative thicknesses of fibres from coarse- and fine-woolled sheep.

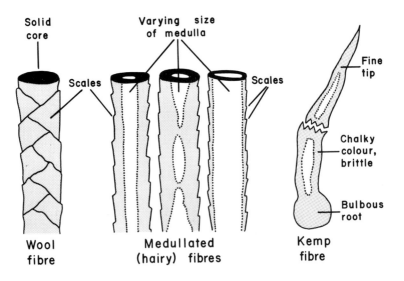

Fig. 6.2 Simplified drawing of different types of fibres.

It follows from the above that the flockmaster who wishes to make the most of his wool clip must bear all these factors in mind but, as with most farm produce, the yield is the most important especially when sheep are kept primarily for wool. Yields vary over a wide range from about 1.3 kg for a Welsh ewe to 2.6 kg for a Suffolk and 5 kg for an English Leicester.

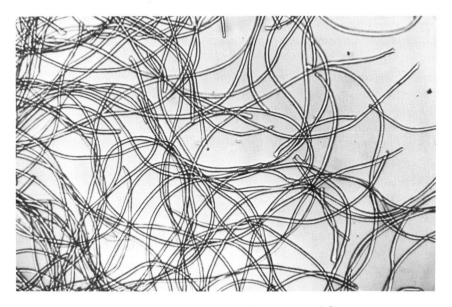

Fig. 6.3(a) Photomicrograph of Merino wool fibres.

Fig. 6.3(b) Photomicrograph of Scottish Blackface wool fibre.

The future of wool

Throughout the present century wool has come under heavy competition from man-made fibres and, in consequence, there have been wide fluctuations in wool prices and profitability. In some instances wool fell to such a low price that, after considering shearing costs, some farmers have speculated that the day could arrive when wool on low ground sheep became nothing more than an embarrassment to be eliminated. The feeling was that sheep might eventually be better employed growing protein that people could eat rather than protein they refused to wear. Had this conception become a reality wool could ultimately have been eliminated by using Wiltshire Horn rams as this breed does not grow fleeces.

The reason why there is still a significant demand for wool is that it has a number of special properties that cannot be matched by synthetic fibres. Wool is very resilient and this springiness allows it to return to its previous form after distortion. It also retains entrapped air in such materials as blanketing, suitings and knitwear to a greater degree than synthetics, giving it an advantage in insulating properties.

Another useful property is that, on being wetted, it absorbs moisture and generates heat. This means that woollen garments worn next to the skin absorb sweat and have a buffering effect on temperature fluctuations.

Finally, a most useful attribute from the standpoint of utilisation is the property wool fibres possess of felting together. The physical structure of wool fibre is such that the scales of the fibre project away from the base of the fibre giving the appearance of stacked flower pots. These scale edges enable the fibres to interlock, thus causing felting (see Fig. 6.2). The fibres of wool adhere together in a way that no other fibres do. This gives great strength to wool-based cloths and also allows very short fibres to be used in manufacturing processes.

These unique qualities suggest that it is unlikely ever to be displaced completely by synthetics. This is especially so in the case of Merino wool. Additionally, as the supply of raw materials (mainly from oil) for the manufacture of synthetics declines and the cost of oil rises the economic position of wool should strengthen further.

Shearing

As the western farming world has evolved sheep that grow wool continuously and do not shed it, arrangements have to be made to remove it at appropriate times in an expeditious and economical manner. The idea, once commonly accepted in Britain and many other countries, that shearing was appointed by nature to be carried out once a year in the early summer, has been substantially modified.

In the higher latitudes where wool growth is slowed down during winter, the fleece is much more readily removed when the regrowth starts in earnest and the wool has, what is known as, 'risen'. This rise of yolk in the wool is accelerated by warm weather and good feeding and is held back by cold, late springs, shortage of food, and stress such as lactation. (The 'yolk' is a mixture of dried sweat — 'suint' — and grease from the sebaceous glands.) The demands made by lactating are underlined by the fact that the barren ewes come to the shearing substantially earlier than do the milk ewes.

Whilst the majority of sheep in Britain and many other parts of the world are shorn in the spring, shearing dates have been varied in different countries in order to bring about an improvement in some other aspect of sheep husbandry. An instance of this is the shearing of ewes shortly before lambing in some parts of New Zealand. The object of the exercise is to eliminate breaks and 'cots' (i.e. lumps of wool which have matted and felted on the sheep's back) in the wool as far as is possible, and this is usually achieved. It is also reported to improve lambing performance due, in all probability, to fewer ewes being 'cast' (i.e. rolling over on their backs and unable to regain their feet) and to their seeking a sheltered place in which to lamb.

In some Australian situations where the wool crop is of major importance and its quality something which must not be jeopardised, the avoidance of ripe seeds and burrs in the fleece may determine that the sheep be shorn before the seeds and burrs ripen.

Animals may be shorn more than once a year. In Britain shearing of lambs has not been common except for some flocks in the south of England, but in Australia lambs are commonly shorn at the ewe shearing while in New Zealand they are shorn after weaning.

In some Scandinavian countries it is becoming the practice to shear ewes before they are housed for the winter and again in early summer. Shearing ewes before housing has been reported as giving heavier lambs at birth than from comparable unshorn ewes. Shorn ewes are under less heat stress as can be seen from their lower respiration rate. In Britain, however, if winter shearing is carried out the shearing is not repeated later as with twice-yearly shearings heavy penalties are imposed by the Wool Marketing Board for partly grown fleeces.

THE ORGANISING OF SHEARING

The flockmaster must make a decision on when to shear well in advance of the day. On hill farms in a country like Britain and on sheep stations in Australia and New Zealand there will be a local pattern of shearing practice. In the days of mixed farming with relatively small flocks, the shepherd or the farmer assisted by farm hands worked their way through the flock over a period of days. On the larger hill farms neighbouring was the common way. The shepherds from a number of farms came together, sheared the sheep of one particular farm and then moved on to another.

In countries such as Australia, South America and New Zealand, with vast numbers of sheep relative to the labour force on individual holdings the professional shearing gang developed. In addition to sheep numbers, Australia, with its vast area and variations in climate, made for a lengthy shearing season which enabled gangs to keep together in employment over quite a long period. Conditions are much different in a small country like Britain. Nevertheless, contract shearing has made great progress over the past twenty years. The advent of the New Zealander, Godfrey Bowen, on the British scene with his demonstrations and schooling of prospective shearers has done a lot to encourage young men to take up the job and good contract shearers are now available in most districts. The Bowen technique is now widespread.

For those who wish to do their own shearing in Britain, the Agricultural Training Board runs classes where sound instruction can be obtained. The big problem in an area of small flocks is that if one is confined to one's own flock shearing has no sooner started than it finishes and all is over for another year. As a number of young farmers have discovered, the thing to do if you have an enthusiasm for shearing is to shear your own flock and then seek a neighbour's on which to further your experience.

PRACTICAL POINTS IN SHEARING

As previously suggested, the original method of harvesting the wool was to pluck it from the live sheep. This in turn gave way to clipping with a pair of hand shears or blades and this method held sway for centuries. Blades have now given place to mechanical or electrically-driven power shears.

Research into chemical 'shearing' whereby animals are dosed with a chemical which interferes with the cell division in the follicle and causes a break in the fibre just below skin level has been going on for some years. Success using thallium was reported in the 1930s and Terrill at Beltsville, Maryland, has had satisfactory results from the use of cyclophosphamids.

While such a method of wool removal could be a boon to the small scale sheep keeper, problems of large scale use are not difficult to foresee. It is unlikely that the period of days which must elapse between the administration of the drug and the time when the fleece could be hand-stripped by the shepherd would be the same for all sheep. This being so the problem of holding the sheep in yards or buildings or running the risk of losing much wool through being pulled off by fences and bushes would have to be faced. A minor problem could be one of sunburn as judging from photographs of treated sheep they are left very bare! A more serious objection to the practice arises from the fact that the ultimate destination of all sheep is the slaughterhouse and one cannot see any public health authority waxing enthusiastic about such treatment of potential human food. It would therefore seem that for the foreseeable future the farmer will need to arrange to have his flock mechanically shorn.

It is not proposed to go into details on the practice of shearing other than to set out what needs to be done and what must be avoided in the clipping of sheep. Shearing is essentially a practical operation which can be mastered only by practice and instruction from an expert shearer in the early stages is a great advantage. One point the apprentice shearer will soon appreciate is that the ability to hold and control the sheep without the expenditure of vast energy is as important as a dexterous manipulation of the shears and handpiece.

There are a number of points which require attention before the actual shearing and the first is whether or not the sheep are to be washed. Washing takes out some of the grease and so lightens the fleece, but washed wool can bring a higher price than unwashed. Some farmers would contend that the increased price never compensates for the loss in weight.

In areas where sheep can contaminate their wool by taking dust baths in red earth or rubbing against peat banks, washing may be necessary in order to avoid being penalised by the buyers for presenting dirty fleeces.

THE OPERATION OF SHEARING

It is often difficult to arrange a shearing so that every aspect of the situation favours a satisfactory outcome. In maritime areas the big problem is to have the sheep dry. Ideally, sheep should be shorn on a dull, warm day and not on a day of cold, sleety showers or under a hot, blistering sun. Also, for the sake of the wellbeing of the sheep, they should be fasted since handling a ruminant with a full paunch can result in the animal dying of cardiac arrest. The sheep must obviously be dry as once a fleece is removed from the animal it is difficult to dry and if packed damp will rot in the bale.

In high rainfall areas it is of great advantage to have a covered area in which to hold sheep overnight prior to shearing. The best type of area has a slatted floor which prevents contamination. Under no circumstances should the holding area be bedded down with straw, peat moss or wood chips, as all are serious contaminants of wool.

The sheep can be shorn on benches or on the floor, the choice depending on the size of the sheep and local custom. The length of 'stubble' left on the sheep could readily be adjusted when the sheep were sheared with blades. The modern powered shearing head cuts quite closely but head pieces known to New Zealanders as snow combs can be used on hill sheep to leave a protective covering. Despite the introduction of snow combs, blades are still used on the high country Merinos and this also applies in highland Britain. The best shearing floor is made of hardwood, as it is usually splinter-free. If a good clean floor is not available the shearing should be done on a tarpaulin, which should be swept regularly.

The job of the shearer is to cut the wool once and the sheep not at all. Double cut reduces the value of the wool as the short bits have to be combed out by the wool manufacturers. It is very easy to damage external organs, the sheaths of rams and the teats of gimmers being particularly vulnerable. This is especially the case when using powered clippers and a warning must be given against carelessness in their use: the fact that they look like overgrown barbers' implements encourages the idea that they are harmless and foolproof. Such is not the case and not only wounded sheep but the gashed thighs of inexperienced and slap-dash shearers point to the lesson of exercising care at all times.

The minor cuts that do occur even at the hands of the careful should be treated with a mild antiseptic. On no account should raw disinfectant or sheep-dip be used.

Having removed the fleece in a clean and dry condition care should be taken to see that it remains clean and dry. Duggings and dirty skirt wool should be removed and packed separately, the fleeces being rolled immediately after shearing. In Britain, the longwool, down and cross fleeces are rolled skin side out, the carpet wools of the hill breeds such as Scottish Blackface and Swaledale, skin side in. When wrapping the fleece the longitudinal sides are turned inwards and wrapping starts from the tail end. The wool is rolled up to the neck and a rope is spun from the neck wool with which the fleece is tied. The wool must never be tied with baler twine or any fibre that can act as a contaminant. Paper twine or waxed string can be used. By the same token sheep should never be marked other than with scourable marking fluid. Branding with pitch or paint leaves a mark which cannot be scoured out. The 'bloom' dipping of tups is another practice which should never take place. The wool of hoggets, ewes and rams should be kept separate from each other.

When wrapped the fleeces should be packed in the wool sheets and stored in a cool, dry atmosphere in a place where there is no danger of its getting wet. In Britain, where all wool goes to registered wool merchants, there is little advantage to be gained from holding it, although the Wool Marketing Board will pay interest on wool held on its behalf. Generally, wool should be despatched from the farm as soon as is practicable.

In conclusion something should be said about the rate at which sheep can be shorn. This, of course, varies with the type of sheep; the difference in the effort needed to handle 100 kg Lincoln rams against 35 kg Welsh mountain ewes is readily apparent. Speed also depends on the state of the fleece. A well risen fleece comes off much more readily than does one where there is no rise, a risen fleece being one in which the growth of fibres has started and the yolk is rising up through the fleece. Finally, the rate of throughput depends on the facilities available and how much work extraneous to shearing the shearer has to do. If someone catches the sheep and presents it to him, gathers up and wraps the fleece and moves on the shorn sheep, a shearer of the Bowen school will get through 300 or more sheep in a day. On the other hand, for farmers and shepherds whose opportunity for practice is limited, 70 to 80 is probably a realistic figure.

Dipping, spraying and jetting

Dipping is a practice designed to get rid of external parasites and cannot be undertaken usefully until some weeks after shearing when there will be sufficient wool length to hold the dip. In many countries the major pest against which dipping is directed is the sheep scab mite; the eradication of this pest requires total immersion. Flock masters should familiarise themselves with the dipping regulations in force in their own particular countries and place themselves in a position to carry them out. It is not necessary to set out here a series of recipes and plans for building sheep dippers. Advice on such matters can be obtained from government advisory services, agricultural colleges and the like and from these the recipient will have the advantage of advice from persons with local knowledge. We will therefore confine ourselves to the important points common to all dippers and dipping processes.

DIPPERS AND THEIR REQUIREMENTS

Dippers may be rectangular in plan, either long or short, where the operator stands at the side of the bath to plunge the sheep or they may be circular where the standing position of the operator is on a central island. The dipper may be a large permanent erection such as a long swim tank associated with a series of holding, forcing and drainage pens together with a shedding race, foot bath and other flock handling equipment. On the other hand it can be a small galvanised or fibreglass portable tank associated with temporary hurdle fencing.

A recently introduced method of total immersion known as 'dunking' is proving satisfactory. Using this system the sheep are placed in a cage with a wire mesh floor and lowered into a tank containing the dip. The cage normally holds about fifteen adult sheep or twenty or so lambs. The tank can be made of concrete let into the ground and forming an integral part of a system of holding and dripping pens, or it can be made of metal and transportable. In the second case, the tank is placed on the ground while a ramp provides access to the cage and a further ramp allows the sheep to leave the cage *en route* to temporary dripping pens. Power for the operation is provided by a tractor fork lift or similar machinery.

The main consideration in selecting a dipper is to ensure that it is appropriate to the flock. With a small flock used to graze very dispersed pastures a portable outfit will probably be the most

Fig. 6.4 A small plunge dipper *(available from C.S.J. & A.M. Sparkes (Development) Ltd, Chilton Polden, Somerset, UK).*

economical. In other cases a centrally placed permanent site with a suitable tank will be more appropriate. Here again the size of flock should control the size of dipper. Too large a tank will be wasteful of time, labour and dip.

A general guide for plunge dippers is that 2–3 litres (about ½ gallon) of made up dip is required per sheep with a minimum capacity of about 650 litres (140 gallons).

The dipping site must provide an all-the-year-round water supply, preferably piped, and the site should have easy access by wheeled vehicles. The above suggestions may strike low ground farmers as peculiar but in many highland areas a fortnight without rain can produce a severe water shortage while a fortnight of incessant rain can make a normally satisfactory site inaccessible even to tractors.

With the above being borne in mind, a dry sheltered position should be sought for the dipper especially if the site is to be permanent. The best position is on a gentle slope which allows for the easy construction of dripping pens. It also allows for pens, races and dipper to be so arranged that sheep are always worked uphill towards the light. Sheep might almost be described as phototrophic animals; their objection to being driven from the light towards darkness will be obvious to anyone who has tried to persuade sheep to go from a bright, sunlit yard through a dark doorway into a building. The site must also be such that the spent dip can

Fig. 6.5 A small plunge dipper and handling pens in action (*available from C.S.J. & A.M. Sparkes (Development) Ltd, Chilton Polden, Somerset, UK*).

readily be disposed of. A soak-away sump is the best method.

That the holding pens should be of adequate size for the flock would seem to be stating the obvious but there are many sheep farms where the obvious has gone unnoticed. The pens should be arranged in such a manner that the sheep can readily be moved forwards and it is an advantage with most plunge dippers to have the pen from which the dipper is entered circular with a diameter of about five metres. This pen is fitted with sweep gates pivoted in the centre; by this means the sheep are emptied from the pen without recourse to manhandling.

The drainage pens should be of such size that the sheep have adequate time to drip without holding up the dipping process. This is particularly important where sheep in full fleece are being dipped as, for instance, against ticks. A Scottish Blackface or Swaledale in such condition removes a large quantity of dip from the tank when it emerges. The dripping pens should have grooved concrete floors with the grooves directed to drain back into the tank via a filter. Finally there must be a satisfactory, safe method of disposing of the spent dip effluent. It must not, under any circumstances, be diverted into a natural water course. Most, if not all, sheep dips are poisonous to fish and other forms of aquatic life and apart from the effect of river pollution on wild life in countries like Scotland and New Zealand where trout and salmon fishing is particularly important, the consequences could have serious financial repercussions on the persons responsible. A soak-away sump situated well away from any streams is quite effective and is usually the most satisfactory method of disposal.

THE DIPPING PROCESS

The first point to check is that the tank is clean and free from sludge in the bottom. The dipper is then filled with clean water and the dip mixed in accordance with the makers' instructions. It must be kept up to strength throughout the operation. This is achieved by the use of a calibrated dip stick. The stick is calibrated by notching the stick at intervals to represent appropriate quantities of dip, say 20–50 litres (5–10 gallons). The quantities must obviously be related to the size of the bath. As the dipping process proceeds and the dip is removed the loss is made good by adding water to which the appropriate quantity of dip has been added. In addition to maintaining the strength and level of the dip every effort should be made to keep it clean as dirt can react with the dip and render it useless. Additionally, sheep can be so contaminated with dirt as to render the process useless if not positively harmful.

As previously stated the wool of the sheep needs to be long enough – about three centimetres as a minimum – to hold the dip. Modern dips contain very efficient stickers which enable the active chemicals to adhere to the wool over a long period. This enables the dip to kill any external parasites the sheep is carrying and also to provide protection from attack for some weeks after the operation. This means that parasites hatching from eggs laid before the dipping are destroyed and this is particularly important in the case of fly strike.

In common with all operations carried out on sheep the animals are best fasted before dipping. They should also be rested and not plunged into a bath in a hot, breathless condition. They must remain in the dipper long enough for the dip to penetrate to the skin and especially in the case of dipping against sheep scab mite the heads must be totally immersed.

When ewes and lambs have to be dipped together it is essential to split the animals into small groups and to dip them as groups. This is to ensure that the lambs are readily mothered up. If the dipping of ewes and lambs is done in large bunches, especially if the dipping is carried out far into the day, a certain amount of mis-mothering is inevitable. A substantial part of ewe-lamb recognition is by smell and as all ewes and lambs will smell equally strongly of B.H.C. or whatever substance was used their natural odour will have been totally disguised for the time being.

The makers' instructions as to how the dip should be handled must be carried out to the letter. Modern organo-phosphorus dips are dangerous and careless use can cause nasty accidents. Sheep going for slaughter in the near future must not be dipped.

SPRAYING

This is an alternative to plunge dipping. It has a number of advantages but in some countries such as Britain it is not approved as a method of sheep scab mite control.

Spraying can be carried out in a spray race or a circular enclosed spray pen. The spray pen has a perforated pipe or pipes above the sheep with other pipes below and to the sides of the race. The spray fluid is forced through the system and produces a heavy spray through which the sheep are run. Equipment for this treatment is manufactured by various companies and the construction of each outfit is such that it can be run from the power take-off of an agricultural tractor.

A variation of the spray race is the spray tank or shower bath. A circular tank with jets above and below can be set up on a permanent or temporary basis. Power for the pressure system is again provided by a tractor power take-off.

The spray race has the advantage over the plunge dipper in that large numbers of sheep can be treated in a short space of time and with the employment of little labour. Upwards of six hundred sheep per hour has been claimed for this method of treatment. The system also has the advantage that there is less contamination of the dip than under the immersion system. Additionally, a smaller quantity of dip is needed in the circulation at any one time although the actual amount of dip carried away by the sheep after dipping will be the same as for plunge dipping. Finally, there is no manual manipulation of the sheep, hence less risk of injury and, where pregnant ewes are concerned, this is very important.

It is usually best, as previously pointed out, to work sheep towards the light but in the case of spray races they should not point directly towards a light source: if the tunnel egress points towards the sun the sheep are confronted with a white wall of spray which makes them frightened and reluctant to move forward.

The total cost of setting up such a system can be more than for a fixed immersion dipper but it is necessary to take a long-term view in judging the economics of the various systems.

The shower system has much in common with the spray race but it takes longer as the sheep have to be held for a period while they are saturated. They are then released and the shower bath refilled with sheep. Where a small flock of sheep is concerned they can be sprayed quite effectively by using a hand lance attached by a rubber hose pipe to a normal farm crop sprayer. The sheep are gathered on a concrete apron or in a corner of a field and the operator clad in rubber boots and oilskins gets in amongst them making passes over and under the sheep with the lance.

JETTING

This technique has been extensively developed in Australia and is now in wide use there. It is carried out with a jetting gun which is, in effect, a short piece of metal tube with four needle jets attached in line. It is operated by a trigger. The jetting fluid is supplied at a pressure of $400-550 \text{ kN/m}^2$ (60–80 lbs per sq. inch) from a medium to high volume crop sprayer. All that is needed for the operation, in addition to the jetting gun, is a piece of hose with which to attach it to the boom connector. The instrument is operated by combing through the fleece of the animal while, at the same time, putting pressure on the trigger of the gun. The advantages of jetting are that a clean wash of constant concentration is applied. There is no waste as the trigger is only operated when the gun is in use. The insecticide is thoroughly applied to those parts of the sheep which are most susceptible to fly strike. Also, particular attention can be given to individual sheep such as a scouring animal. Jetting is a quite speedy operation and over 100 sheep per hour can be treated with one gun.

A final warning must be given on spraying of all kinds: the chemicals in the dip should be in solution. Dips which carry constituents in suspension can cause trouble as the suspensions tend to clog the nozzles.

SELECTING A SYSTEM

It will be apparent from the above that the methods and equipment used against external parasites of sheep vary extensively and the flockmaster should select and adapt a system appropriate to his particular farm, types of sheep and the nature of his parasite problems.

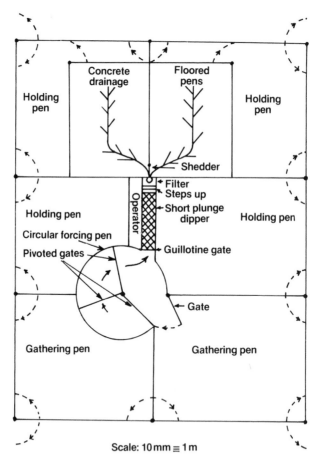

Fig. 6.6 Example of dipping arrangements using a short plunge dipper.

In order to achieve a satisfactory set-up the farmer should consult with other farmers and shepherds operating under similar conditions and examine systems that have given satisfaction over a number of seasons. From his observations and the advice he receives the farmer should construct a system for himself which is a synthesis of the good points obtained from others. This approach is much better than selecting a system and set of plans from a text book or other source. The systems which are operationally the best are rarely obtained 'off the peg'. Fig. 6.6 shows the basic requirements of a dipping set-up.

The discussion on dipping and spraying virtually concludes the round of chores associated with sheep keeping and the shepherd's year and it is now time to turn to production systems.

Before turning to systems, however, it is necessary to take a brief look at the sort of product the market will accept as to produce an article first and then go in search of a market afterwards is not really a sensible economic exercise. Our next chapter will therefore deal with factors affecting carcase composition, meat quality and market requirements.

Points to remember

1. Factors affecting the quality of wool:

   Fineness.
   Uniformity.
   Suitable length.
   Elasticity and strength.
   Crimp.
   Colour.
   Lustre.
   Behaviour to dyes.
   Freedom from kemp.

2. Points of importance in shearing:

   The sheep must be dry, fasted and with a clean fleece.
   Shear on a clean surface and keep clean.
   The wool must be cut once only, fleeces wrapped immediately and tied with wool or plastic cord, sheeted and stored in the dry.

3. Dipping and spraying:

   The dip or spray must be kept at the recommended concentration and kept clean.

The sheep must be rested, carry sufficient wool to hold the dip and immersed for an appropriate period.

Because mothering up is difficult, ewes and lambs should be dipped in small batches only and not late in the day.

Dip must be safely disposed of.

7 Meat quality and market acceptability

The origin of almost all agriculture lies in subsistence farming where the farmer satisfied as many of his own and his family's needs as possible from his own land. Under such a system any surplus production was disposed of by sale or barter. Taxes to an overlord or tithes to a church would also have to be satisfied before any surplus accrued.

In spite of the fact that in Western Europe there has been a market of sorts since the Roman era the goal of self-sufficiency has been followed by many up to recent times. Fifty or sixty years ago farmers in Britain still kept one or two house cows to provide milk and butter, killed a couple of pigs a year for bacon, ran a barnyard flock of hens and grew an acre or so of potatoes for household use.

It is also worth remembering that a hundred years or so ago the homesteaders opening up the virgin lands in the Americas, Australia, New Zealand and South Africa also aimed for self-sufficiency as an important goal.

The reason for giving this historical reminder is that the idea of surplus sales still lingers amongst some farmers in that they tend to produce a commodity, take it to market and expect the market to absorb it. This is, of course, more true of Europe than, say, New Zealand. Modern farmers need to be market-orientated and should examine the market with a view to learning its requirements and devising an economically satisfactory production system to meet them. This is not to say that farmers should reject advertising as a means of persuading the public of the merits of what they wish to sell.

In respect of meat, a large majority of housewives present a negative attitude rather than a positive one. They know what they do not want. The meat must not be tough or overfat or have too much bone or be difficult to cook. Large joints are not required except by the catering trade. The above provisos apply to all meats but, in addition, sheep meat has other attributes about which the producer can do nothing. In most European countries and North America, sheep meat is much less favoured than beef and pork. One reason is the high melting point of the fat, which often results in it congealing on the plate and on the palate. The other is that many people appear not to like the smell of roasting mutton.

The farmer therefore needs to produce an article which is tender, succulent, low in fat and bone, and not strongly flavoured — in other words, a carcase from an animal which is relatively young and small. In order to produce such an animal we must now look at the factors which affect the quality of a carcase.

Factors affecting carcase quality

AGE

The animal must be of such an age at slaughter that the edible joints are well developed, but this will also depend on nutrition and on the breed or cross of the animal. The age at slaughter may be from eight to ten weeks for an 'Easter lamb' and upwards. The flesh of a young animal has a much higher moisture content than that of an older animal, and while this makes it more succulent it also results in greater shrinkage on roasting. As the animal advances in age the muscle fibres coarsen, the sinews and connective tissues develop, the meat gets tougher, the colour of the flesh deepens and the flavour gets stronger. In Western Europe older animals are unacceptable for cooking as joints and go for manufacture.

SEX

Males grow faster than females and have a larger mature size. This means that male lambs can be taken to higher weights before becoming over-fat than can females. Well-nourished ewe lambs readily become too fat for modern requirements. With age, rams become coarse-fleshed and strongly flavoured much more so than do ewes. In consequence their carcase value is low. Castrates are intermediate in development between males and females. As well as their secondary sexual characteristics (such as horn growth) being depressed, they put on more fat and grow more slowly than do entire males.

BREED OR TYPE

In the improved breeds of sheep the fat is laid down subcutaneously and between the muscle fibres, giving rise to 'marbling'. In unimproved breeds it is laid down mainly around the gut and kidneys and, in some eastern breeds, around the tail. The improved mutton types have shorter, thicker and heavier bones than their unimproved cousins. They are shorter in the leg and blockier in conformation, the aim of the breeders being to obtain an animal which is heavy and full-fleshed in the hind legs and loin as these are the most valuable joints. These improved animals are early-maturing; in other words, they reach their slaughter weight much sooner than do unimproved animals. This always assumes that their nutrition has been satisfactory.

NUTRITION

This is of paramount importance as, whatever the potential of the animal, the potential weight for age will not be achieved if the food supply is short or of poor quality. Under conditions of poor nutrition an improved animal will not perform any better than an unimproved one, probably worse.

CONDITION

This depends on the fat cover on the carcase. A joint requires a sufficient cover of fat in order that it does not dry out on roasting. Too much fat has either to be removed by the butcher or cook, or left by the side of the plate. This is a serious waste as fat is much more costly to put on an animal than is lean meat. The objective is therefore to produce lamb carcases that have just enough fat to ensure that (a) they roast well and (b) minimum shrinkage of the carcase occurs while it hangs in the butcher's shop.

EXERCISE AND METHODS OF SLAUGHTER

The more exercise an animal has the better is the development of sinews, tendons and muscles. In other words, it gets tougher. Also violent exercise and excitement immediately before slaughter make the animal difficult to bleed. Rough handling can give rise to bruising which results in blemished carcases and can also cause dehydration.

Sheep, like any other animals, should be fasted before slaughter. They should be in a rested condition and, under no circumstances, should any animal be running a temperature as this also interferes with bleeding. Finally, after slaughter, the carcase should be chilled and hung at a temperature of around 0°C for a few days.

The carcase

THE ATTRIBUTES OF THE ACCEPTABLE CARCASE

All the above factors add up to the ideal of a relatively small carcase from a young animal which is lightly covered with fat. Putting weights and percentages to such carcases give the following ranges of acceptability, at least for the British market. This market is the largest in the world for lamb carcases and, until recently, over half came from overseas, principally from New Zealand. These New Zealand carcases were of light weight, relatively lean and strictly graded, and set the standard, especially for the South of England and London area where the domestic market is for a lamb of 13–16 kg (28–35 lbs) in weight. Elsewhere in Britain the market demands lamb weighing 14–18 kg (30–40 lbs) for domestic customers. The wholesale catering trade will accept 18–23 kg (40–50 lbs) carcases, and the maximum for hoggets is about 27 kg (60 lbs) for the meat processing industry.

DRESSING OUT PERCENTAGES

As with all ruminant animals the dressing out percentage (i.e. the weight of the carcase expressed as a percentage of the total original body weight) is low compared with swine. It can be as low as 40% for a light-weight animal in poor condition. Meat lambs vary from 45–50%, finished hoggets from 47–53% while a fat, heavy sheep may be up to 60%. These percentages apply to sheep that have been dressed in the British manner, namely having all the internal organs (other than the kidneys) removed together with the skin, head and feet. The most valuable part of the sheep's offal is the skin; in Britain, skin wool provides about 40% of home-produced wool. Of the rest of the offal, the liver is the most important and valuable part.

CUTS OF LAMB

The cuts into which lamb carcases are divided vary from country to country and even from area to area in any one country. Generally speaking, lamb is not boned out but sold in joints, although shoulders may be boned and rolled. The number of commercial cuts from a carcase depends in the main on the size of the carcase and local butchering practice. While small milk lambs may only be quartered, large hoggets may provide a dozen cuts from a side.

The common English cuts are shown in Fig. 7.1, and the bone structure of the carcase in Fig. 7.2.

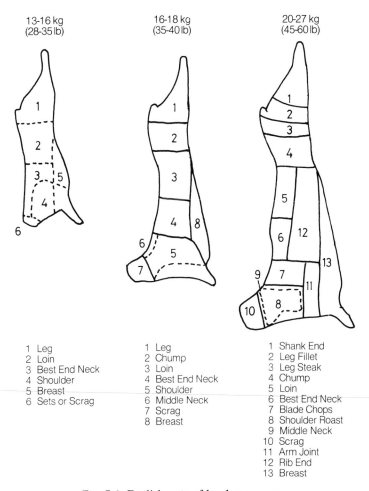

13-16 kg
(28-35 lb)

16-18 kg
(35-40 lb)

20-27 kg
(45-60 lb)

1 Leg
2 Loin
3 Best End Neck
4 Shoulder
5 Breast
6 Sets or Scrag

1 Leg
2 Chump
3 Loin
4 Best End Neck
5 Shoulder
6 Middle Neck
7 Scrag
8 Breast

1 Shank End
2 Leg Fillet
3 Leg Steak
4 Chump
5 Loin
6 Best End Neck
7 Blade Chops
8 Shoulder Roast
9 Middle Neck
10 Scrag
11 Arm Joint
12 Rib End
13 Breast

Fig. 7.1 English cuts of lamb carcases.

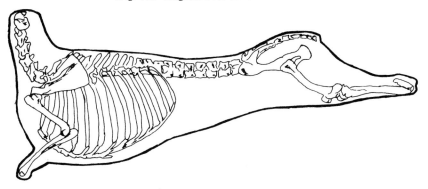

Fig. 7.2 Carcase of sheep showing bone structure.

THE MEAT AND LIVESTOCK COMMISSION CARCASE CLASSIFICATION
SCHEME

We have already commented on the strict grading of New Zealand
lamb for the British market. In Great Britain, the Meat and Livestock Commission have introduced a comparable classification
system for the home-produced animal which consists of 5 'fat
classes' according to the amount of fat, and 5 categories of 'conformation' or distribution of meat on the carcase. Carcases are
classified by visual examination only, unlike New Zealand where
an instrument for measuring the depth of fat is used (see Fig. 7.3,
Fig. 7.4 and Fig. 7.5). Conformation is discussed in the next section.

			Fat class		
		1 (very lean)	2 2	3 3	4 4	5 (very fat)
Conformation	Extra	1E	2E	3E	4E	
	Average	1	2	3	4	5
	Poor			C		
	Very poor			Z		

Fig. 7.3 Sheep Carcase Classification Scheme (*Meat and Livestock Commission*).

Selecting lambs for the market

The farmer needs to make himself familiar with what the market
requires and with the basis of current classifications. He then needs
to relate it to the live animal and have a set of criteria by which he
can form some judgment of how the live animal will appear as a
carcase. This requires experience of handling animals before slaughter
and then seeing them when they have been slaughtered and classified
into their appropriate groups.

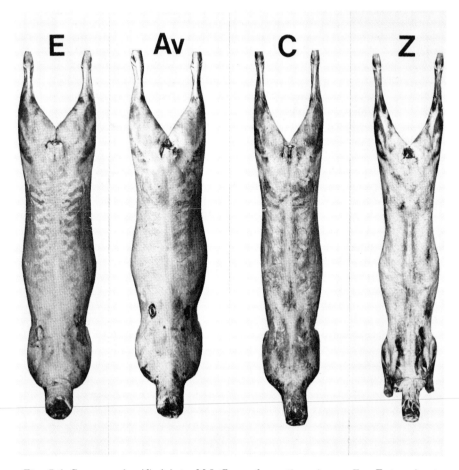

Fig. 7.4 Carcases classified into M.L.C. conformation classes. E = Extra; Av = Average; C = Poor; Z = Very poor. (*Reproduced by permission of the Meat and Livestock Commission.*)

The first requirement in producing an acceptable meat lamb is a breed or cross of sheep which, mated with a suitable ram, gives a lamb which when properly fed meets the above criteria. In other words, it is a fast-growing lamb of good *conformation*. The Meat and Livestock Commission defines conformation thus: 'lambs with good conformation have thick loins, compact shoulders and thick, round legs'. On the other hand, lambs with poor conformation have narrow, thinly fleshed shoulders and loins with thin legs lacking in 'second thigh'. It is the possession of this second thigh or twist which is one of the important contributions of the Southdown to modern mutton breeds.

FAT CLASS

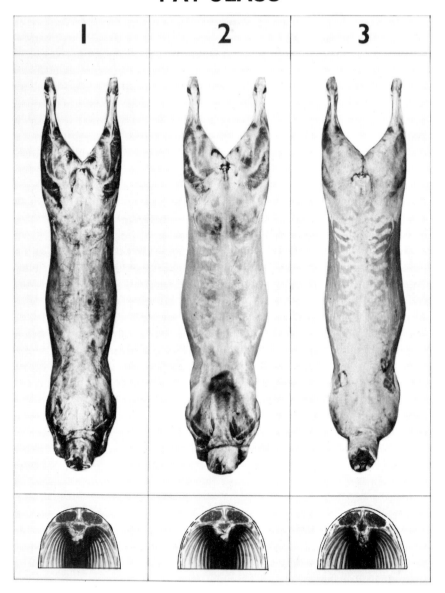

Fig. 7.5 Carcases showing
M.L.C. fat classification.
(*Reproduced by permission
of the Meat and Livestock
Commission.*)

FAT CLASS

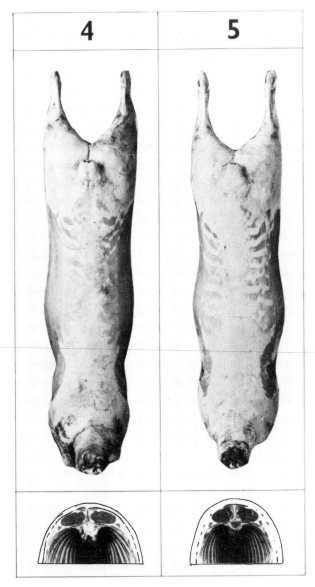

Earlier in the book a Meat and Livestock Commission scheme for evaluating the physical condition of breeding ewes was described and M.L.C. have produced a similar scheme for assessing the marketability of lamb. The system is, naturally, related to their carcase classification scheme and is as follows.

There are four areas of the live animal shown in the diagram (Fig. 7.6) which can be handled with advantage to determine the fatness of the lambs and these points provide a guide to the resulting fat class of the carcase.

A. Around the tail root (dock).
B. Along the spinous processes of the back bone and over the eye muscle and the tips of the transverse processes in the lumbar region.
C. Along the spinous processes of the back bone over the shoulder.
D. Along the breast bone (sternum).

Of these four points A and B are the most important and generally used.

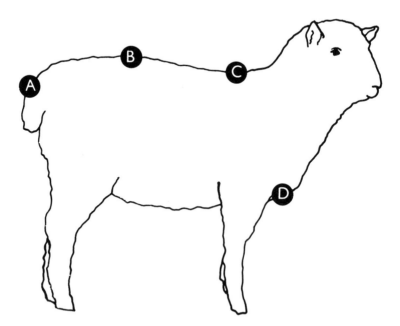

Fig. 7.6 Handling procedure recommended by the Meat and Livestock Commission for assessing the fatness of live lambs. (*Reproduced by permission of the Meat and Livestock Commission.*)

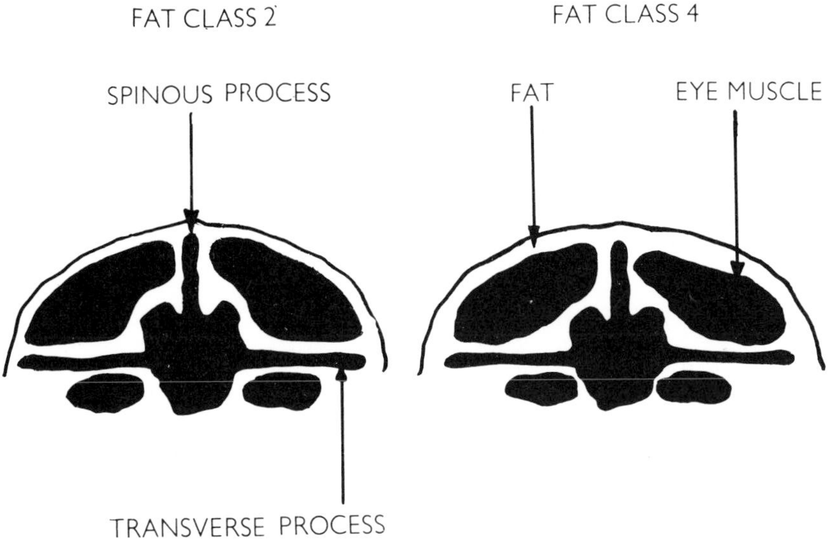

FAT CLASS 2   FAT CLASS 4

SPINOUS PROCESS   FAT   EYE MUSCLE

TRANSVERSE PROCESS

Fig. 7.7 Handling points. The diagrams show cut surfaces of loin (at 3rd/4th lumbar vertebrae) to illustrate fat cover over the eye muscle in carcases of Fat Classes 2 and 4 and show the position of the spinous processes. (*Reproduced by permission of the Meat and Livestock Commission.*)

The degree of fatness at A is assessed by handling the tail root (dock) to see how much fat is covering the bones. As a lamb becomes fatter it is more difficult to detect individual bones (see Fig. 7.7).

The degree of fatness at B is assessed by placing the hand over the transverse processes of the loin to assess the degree of prominence of the processes. The less prominent the processes are, the fatter the lamb is.

In handling the lambs in these examinations it is important to apply the minimum of pressure otherwise bruising may be caused, resulting in a damaged carcase.

The following five point scale enables the live lamb to be related to the Sheep Carcase Classification Scheme:

Fat Class 1
Loin   Spinous processes very prominent. Each process can be readily felt.
        Transverse processes prominent. It is very easy to feel between each process.
Dock   Fat cover thin. Individual bones very easily detected.

Fat Class 2

Loin   Spinous processes prominent. Each process can be felt
        easily.
        Transverse processes: each process can be felt easily.

Dock   Fat cover thin. Individual bones can be detected with light
        pressure.

Fat Class 3

Loin   Spinous processes: tips rounded. With light pressure indi-
        vidual bones can be felt as corrugations.
        Transverse processes: tips rounded. With light pressure
        individual bones can be felt as corrugations.

Dock   Individual bones can be detected with light pressure.

Fat Class 4

Loin   Spinous processes: tips of individual bones can be felt as
        corrugations with moderate pressure.
        Transverse processes: tips can be detected with firm pressure.

Dock   Fat cover quite thick. Individual bones can be detected only
        with firm pressure.

Fat Class 5

Loin   Spinous processes   } Individual bones cannot be detected
        Transverse processes } even with firm pressure.

Dock   Fat cover thick. Individual bones cannot be detected even
        with firm pressure.

Having discussed the type of carcase popular on most western
markets and given a method for deciding what sort of carcase a lamb
is likely to produce, the next consideration must be the factors
which control the production of suitable animals.

Growth and development

The rate of increase in weight and size of the growing animal follows
an S-shaped curve as shown in Fig. 7.8. The rate of growth is rapid
from conception to birth and thence to puberty. At puberty, the rate
of growth starts to slow down until the age of maturity is reached.

The most cursory of glances will show that a new born lamb is
not a miniature of an adult sheep. It appears to be all head and legs.

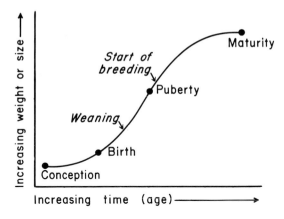

Fig. 7.8 Simplified growth curve.

This means that it does not simply increase in mass as it ages but exhibits differential growth. The tissues of the body attain their maximum growth rates at different times and in fixed chronological order. Hammond and his colleagues at Cambridge worked out these changes in some detail in the 1950s and 1960s. The order of development, not surprisingly, follows the order of use. The lamb, as with other hooved animals, has to be able to 'get up and go' shortly after birth. It is therefore born with a strong set of functional legs. The first growth stage is therefore one of bones and internal organs. This is followed by the development of muscle or lean meat. The final stage is the laying down of fat in quantity.

In the late maturing breeds these phases tend to be distinct from one another. Such animals will lay on fat readily when growth has practically ended but they will not fatten satisfactorily while bone and muscle growth is still active. Early maturing breeds such as the Down breeds have these phases telescoped so that quite young lambs lay on both meat and fat long before the growth of bones and internal organs is completed. In order that early maturing animals may express these qualities a high level of nutrition is essential. On a low plane of nutrition the development of an early maturing breed will follow the pattern of a late maturing animal. It is possible to find breeds of sheep which combine quite good fleshing quality, relatively early maturity and reasonable fecundity from which meat lambs can be produced but the majority of meat lambs are crossbreeds. The sorts of systems which can be used for the production of meat lambs will be discussed in the next chapter but first we must remind ourselves that the purpose of meat lamb production is profit. We must therefore look at the major factors on which profit depends.

Before doing this a few words should be said about a change in nomenclature which may have struck the older reader. What were called 'fat lambs' are now called 'meat lambs'. This reflects the modern abhorrence of fatty meat. A fat lamb is anathema to the butcher today; what he wants is lean meat, not fat. In the search for animals with the propensity for laying down flesh with little fat, breeders are now using ultrasonic instruments for the examination of fat depth in the live animal.

Points to remember

1. Factors affecting meat quality.

> Age.
> Sex (including castrates).
> Breed or type.
> Nutrition.
> Body condition.
> Exercise.
> Method of slaughter.

2. Market acceptability.

> Today's demand is for a young carcase, only lightly covered with fat.

3. The M.L.C. Carcase Classification Scheme.

> The M.L.C. Scheme uses two parameters — fat class and conformation.
> There are 5 classes of fatness from 1 (extremely lean) to 5 (extremely fat) and 4 classes of conformation from 'very poor' to 'extra'.
> Good conformation = thick loins, compact shoulders and thick, round legs.

4. Selection of animals for market.

   For fat class use point scoring on touch at four points:

   (A) at the tail root
   (B) over the lumbar area
   (C) over the spinous process above the shoulder
   (D) along the breast bone.

8 Profitability in sheep farming

This book does not set out to teach the financial side of farm management, and such phrases as gross margin analysis, equi-marginal returns and cash flow will be used as little as possible. Nonetheless, in dealing with systems both students and farmers need to be reminded of the economic consequences of various actions. In such applied sciences as agriculture the individual expert may overemphasise the importance of his own speciality and the farmer must guard against this danger. What has to be grasped is that the main objective of the farmer is *not* to own the healthiest flock of sheep, produce the heaviest lamb crop or sell the best lambs at a particular market, but to produce the best profit consistent with safeguarding his capital assets.

These observations may sound like heresy to the student fresh from the clean-cut concepts of pure science but a few examples will illustrate the point. If, for instance, only one ewe per acre must be run on lowland grass in order to maintain a minimum worm infestation in the sheep, then this enterprise will obviously be an economic failure. If one has a hill flock of ewes that are too 'well fed' at tupping time and then fall on hard times during late pregnancy then twin lambs will be small and weak at birth, the ewes will be short of milk and the situation will be worse than had the ewes conceived only singles. In the case of a meat lamb producing flock, one excellent lamb per ewe is much less profitable than two moderately good ones.

Profitability

The job of the farmer is to ensure that he has an enterprise which exhibits long-term profitability and where his farm shows a steady improvement. In essence, profitability means that the income of the farm is in excess of the expenditure, the best profit being made by a careful control of inputs and a sustained endeavour to increase output. The major factors relating to profitability are discussed below.

THE COST OF LAND AND SALE PRICES OF SHEEP PRODUCTS

In any type of enterprise the most important factor is the sale price of the article produced. If this is too low no amount of expertise in production will compensate for the lack of return. In a given market condition there is a limit to what can be paid for land either in purchase price or rent. Anyone who buys or rents land too dearly relative to the market must ultimately fail. This is heavily underlined by what happened to many British farmers in the late nineteenth and early twentieth centuries. For example, farmers in Britain had made quite a lot of money during World War I and their euphoria about the future caused intense competition for land. The price of land and rents rose sharply after the War but a sudden and serious slump in the early Twenties resulted in the bankruptcies of many farmers.

STOCKING RATE

An examination of the records from a large number of sheep producers indicates that the most important factor in profitability is the *stocking rate*, stocking rate being defined as the total number of sheep divided by the total area of pasture devoted to their support over a given period. It is usually expressed as sheep per hectare or acre, although on some of the harder Scottish hills and the more arid farming areas of Australia hectares per sheep is more appropriate. The term *stocking density* is used to describe the number of sheep carried per unit area in a particular field at a particular time.

When an examination of comparative financial records from a number of sheep farms is made, almost invariably it is the farmers with a high stocking rate that are most successful. This does not mean that overstocked farms are always profitable but that disease control and nutrition should be so arranged as to permit the optimum output per hectare.

DISEASE CONTROL

Disease control is a factor of prime importance especially where one has intensification. Infectious diseases of lambs and also stress diseases of ewes assume greater importance in intensive units than in the less intensive systems of management. It was shown in Chapter 3 that there are a large number of diseases and other afflictions to which sheep are subject but also that it is possible to control these

problems by good management and, in intensive situations, good
management is imperative.

FLOCK DEPRECIATION
A factor in which disease control has a marked influence is flock
depreciation. In the case of self-replenishing flocks such as those
kept under hill and range conditions, heavy losses of ewes or lambs
will have a number of serious consequences. There will be fewer
cast ewes for sale. This, in turn, means more ewe lambs must be
kept for replacements. There will be fewer lambs for sale and also
the amount of selection on the female side that can be practised
will be reduced. In other words, a flock with a high depreciation
rate costs more to maintain and gives a lower return than a sound
flock.

There is also a side effect where lactating ewes are lost. The
shepherd is left with lambs that not only need supplementary feeding and more attention but grow more slowly than their non-orphaned
companions.

SOUND USE OF LABOUR
Intensification of production increases the demands made on skilled
labour and careful thought must be given to such things as the programme at lambing time and the integration of less skilled workers
into the work with the ewes, in order that the time of the shepherd
is concentrated on skilled jobs. The arrangement of the dipper,
handling pens, feeding systems and the like must be such that labour
demands are kept to a minimum.

NUTRITION
The food provided must be satisfactory for the type of production
but it must not be too costly. This applies not only to supplementary
concentrate food but also to pastures. On mixed farms catch crops
such as turnips, rape, kale, Italian rye grass etc. are commonly grown
to supplement pasture and, while these may be very acceptable to
the sheep, there are many cases when they are not economical.

MANAGEMENT POLICY
This, baldly stated, would seem self-apparent but quite often farmers
are to be found working systems which are not appropriate to their
particular farms or the local markets. In many cases these farmers
are technically competent but their profits would be greater with a

change of system. Examples include producing crossbred store lambs by tups which are not locally popular, or running a purebred meat lamb producing flock when a crossbred flock would be more appropriate.

Another fault is lambing at the wrong time – for example, lambing grassland ewes too early to take full advantage of the grass but failing to get most of the lambs to the early spring market, thus pushing up the amount of concentrates needed but failing to sell the lambs at a price commensurate with the costs incurred. The time of lambing must always be arranged so that the food supply and costs are related advantageously to the market for the product.

A COMPARISON OF TOP AND AVERAGE PRODUCERS
The following figures taken from the M.L.C. Commercial Sheep Production Year Book 1980-81 give a concrete example which underlines the economic points which have been made.

Although we do not wish to delve too deeply into the financial side of management it is first necessary to define some of the terms used in order that the information may be better understood.

The normal methods of comparing farming enterprises is by *gross margin analysis*. The gross margin of an enterprise is the output minus the variable costs. The profit of a farm is the sum of the gross margins of each of the farm's enterprises minus the fixed costs. Examples of variable and fixed costs are shown in Table 8.1.

Table 8.1 Examples of costs.

Variable costs	Fixed costs
Purchased and home-grown feeds	Rent
Veterinary and medicines	Labour
Variable costs of grass and forage crops	Machinery depreciation and repairs
Miscellaneous (bedding, transport etc.)	Sundries (telephone, electricity)

The output of a flock is the total returns from lambs and wool, plus any subsidy which may be received, together with the value of any lambs left on hand at the end of the sheep year, minus the cost of flock replacement.

On examining the M.L.C. Lowland Flock Analysis shown in

Table 8.2 M.L.C. Lowland Flock Analysis (taken from the M.L.C. Commercial Sheep Production Year Book 1980–81).
Physical and financial results for 380 lowland flocks selling most of their lambs off grass in summer and autumn 1980.

PHYSICAL RESULTS	Average	Top third
Av. flock size (ewes to ram)	398	387
Ewe to ram ratio	40	41
Ewe lambs in breeding flock (%)	10	11
Per 100 ewes to ram:		
No. of ewes:		
empty	7	5
dead	4	3
lambed	92	93
No. of lambs:		
born dead	10	9
born alive	150	164
dead after birth	9	7
reared	141	157
retained for breeding	6	7
sold finished	88	107
Stocking rate (ewes/ha):		
summer grazing	14.2	16.8
overall grass	12.0	14.6
overall grass and forage	11.5	14.2
N fertiliser per hectare (kg)	151	181
N fertiliser per ewe (kg)	13	12
FINANCIAL RESULTS (£ per ewe)		
Output		
Lamb sales (a)	38.49	43.65
Wool sales	3.15	3.18
Gross returns	41.64	46.83
Less flock replacements	6.45	4.96
Output	35.19	41.87

FINANCIAL RESULTS (contd)	Average	Top third
Variable costs		
Concentrates (b)	6.25	5.80
Purchased forage	0.45	0.47
Fertiliser	3.65	3.88
Other forage costs	0.70	0.95
Total feed and forage	11.05	11.10
Vet and medicines	1.86	1.76
Miscellaneous and transport	0.95	0.78
Total variable costs	13.86	13.64
Gross margin (output − variable costs)	21.33	28.23
Gross margin per grass hectare (£)	258.80	412.16
Gross margin per hectare (£)	246.34	399.51
(a) Av. return per lamb (£)	27.3	27.8
Estimated return per kg lamb carcase (£)	1.52	1.54
(b) Concentrate cost per tonne (£)	117.9	116.0
Ewe concentrates per ewe (kg)	46	44
Lamb concentrates per ewe (kg)	7	6

The top-third flocks achieved a gross margin per hectare £153 (62%) higher than average. Table 8.3 shows the contribution made by different factors to this extra gross margin.

Table 8.2 the reader will probably be struck by the fact that for many of the figures there are only small differences between the top third of producers and the average. For instance, top producers had 5% barren ewes compared with an average 7%; similarly, the differences in perinatal deaths seem slight. It will be noted, however, that *all* the adverse factors are on the same side and the *cumulative* difference in the totals is considerable. The major differences are in the numbers of lambs born and reared and in the stocking rate, where the top farmers have over two ewes more per hectare than the average.

The costs of both groups are practically the same, and it is interesting that although the nitrogenous fertiliser used per hectare is higher for the best performers it is marginally less when stated per ewe.

Table 8.3 shows where the contributions to the superior performers came from. An examination of figures from other flocks in other situations shows similar trends.

Table 8.3 Contribution to top-third superiority in gross margins for lowland flocks, 1980.

	% contribution	
Sale returns per lamb	7	
No of lambs reared	33	Top-third flocks made £3063 extra gross margin per 20 hectares of grass
Flock replacement cost	11	
Feed and forage cost	2	
Stocking rate	37	
Other factors	10	

The two major factors influencing gross margins were the number of lambs reared and the stocking rate. On average, for each increase of 0.1 in the number of lambs reared per ewe the gross margin per ewe increased by £2.56, and for each increase of 0.1 in the number of ewes per hectare the gross margin per hectare rose by £1.89.

Sheep production systems in common use

1. HILL, MOORLAND AND ARID AREAS

As mentioned earlier in this book, sheep from hill, moorland and arid areas form the basis for much of the world's sheep trade. The reason for this is that, of all domesticated animals, they are usually the species best adapted to the exploitation of such places. Possibly, however, red deer kept for slaughter as opposed to sport may one day compete seriously with the sheep in mountainous areas of countries such as Britain and New Zealand. The attraction of venison to people who have marked antipathy to fat is quite strong but the production of high quality wool from sheep such as the Merino should keep the position of sheep secure in the arid areas, while the cost of fencing and lack of good outlets for worn out hinds will probably limit the expansion of deer.

The major characteristic of these areas of extensive stock rearing is that much of it has thin, poor soil which limits production, and the growing season is greatly restricted by long, severe winters as in the Scottish Highlands or drought as in some areas of Australia and South Africa. The sheep used in exploiting such areas are small and active. As Hagedorn pointed out many years ago, two small animals with eight legs and two mouths are much more effective in harvesting scarce material than one large animal with only four legs and one mouth.

The greater number of these hill and similar sheep holdings are such that, due to terrain, climate, water shortage and similar constraints, they are not susceptible to any large scale improvements to the extent that they could be changed over to some other and much more intensive type of agricultural production. Some suitable areas, however, may go over to forestry, nature reserves, national parks and the like. If it is accepted that the above categories of sheep farm will remain with persons happy to farm them, then it follows that a great deal of the meat lamb production in the world will depend on the stratification mentioned earlier in the book. The people who farm such areas do so as much as a way of life as an economic exercise. Indeed, farming as a whole shows personal preference as a determining factor in how the land is farmed to a much greater extent than is seen amongst, say, factory owners and what they produce.

2. GRASSLAND AND MIXED FARMS

The uses that farmers can make of hill, moorland and arid areas are limited. The options are much wider on grassland and mixed farms where farmers have a choice of livestock and a choice of systems. The major portion of the world's sheep production is based on the grazing of grass and other herbage such as legumes and shrubs. Most sheep-keeping farms other than in the hill and marginal areas show some element of mixed farming but the time spent by sheep on these farms on crops other than grass tends to be small. This is certainly the case in Britain today, unlike the eighteenth and nineteenth century when sheep spent much of the year on arable crops, but there may be a swing back to the use of specialist crops for sheep if the rise in oil prices and hence the cost of fertilisers continues. At the present time the majority of sheep being fed on arable crops such as turnips and swedes are hoggets being fed ready for slaughter.

The farmer of a mixed farm has a number of advantages over the all-grass farmer; some of these arise from the fact that a mixed farm implies a rotation of crops. Crop rotations were developed to control weeds and pests, to ameliorate the land with such crops as clover and to provide different food sources for their livestock. Substantial use can also be made of crop residues such as sugar beet tops, Brussels sprouts and other brassica crops.

The rotational system known as ley farming, whereby arable crops are followed by a number of years under grass, provides an excellent basis for sheep farming. The rotation not only breaks up the plant disease and pest cycles but also has a similar effect in controlling sheep stomach worms (see also Chapter 12).

On large mixed farms there can be another bonus: spare labour is available in the early spring to help the shepherd at lambing time, provided the lambing is arranged to fall before the busiest period of the spring arable work.

3. ARABLE SHEEP FARMING

Arable sheep farming in Britain has declined to a vestige of what it was in the early nineteenth century. In the height of its popularity it was to be found in many areas but the strongholds were the drier areas of the East and South of England on either light, sandy soils or the calcareous soils derived from chalk or limestone. These soils are free-draining and consequently are very suitable for sheep as the problem of foot rot tends to be much less in evidence than it is on clay or loam soils. These arable farms fell into two classes, those which had access to rough grazing such as down or heathland and those which did not. The flock was sustained over the years by a series of succulent crops which not only provided food but also water. The crops grown were brassicas such as turnips, swedes, mustard, kale, rape and cabbages, legumes such as red clover, lucerne (alfalfa) and sainfoin and graminaceous crops such as rye and ryegrass.

This consuming of crops in situ and the treading of the hooves of the sheep led to an increased fertility and improved soil structure which made possible the growing of valuable cash crops such as wheat and malting barley from poor, light soils.

The system involved *close folding* of sheep over the ground. Close folding is a concentration of sheep on to a small area of crop relative to the number of animals being grazed and is normally accomplished by the use of hurdles or – nowadays – by electrified netting. Close folding, however, was very expensive in labour and when competition came, first with cheap cereals from the Americas followed by cheap frozen mutton from the Antipodes, it could not survive. The systems with no accommodation grassland suffered most and they were the first to disappear.

In Britain today the flocks which are mainly arable are almost, without exception, ram breeding flocks producing Down rams. This is because ram breeding flocks require the lambs to be born early in the season — December/January — so that they can be used in the first year of life.

REGIONAL DIFFERENCES

While these various systems of sheep production can be found over wide areas there is, nevertheless, a number of broad regional differences. The sheep farming of Britain shows this quite clearly. The north and west, i.e. Scotland and Wales and parts of the West Country, are generally speaking much higher and also wetter than the rest of the country. Consequently, they have the mountain and hill flocks. The high rainfall favours grass production, there is more pasture than arable land, and these areas have a preponderance of breeding flocks. The rest of the country is mainly of much lower elevation, much drier, and has large areas of deeper, richer soil which are consequently better orientated to arable farming. It is to these areas that many of the feeding hoggets from the north are sent to finish on roots, sugar beet tops and the like.

A detached map showing these individual distributions would be very complicated unless the scale was very large. The M.L.C. however, has prepared a map showing the distribution of sheep, numbers of farms and slaughterings for the regions which gives some idea of sheep movements. These are shown in Fig. 8.1.

Figure 8.1 compares the regional distribution of sheep numbers in June 1980 and slaughterings in the year ended 31 March 1981. The three most important sheep producing areas, Wales, Scotland and the North of England, produce many more sheep than they slaughter. Yorkshire and Lancashire and the South West have the greatest net influx of sheep for slaughter. Table 8.4 shows regional sheep numbers for England with those for Wales, Scotland and N. Ireland added later from the June 1980 census.

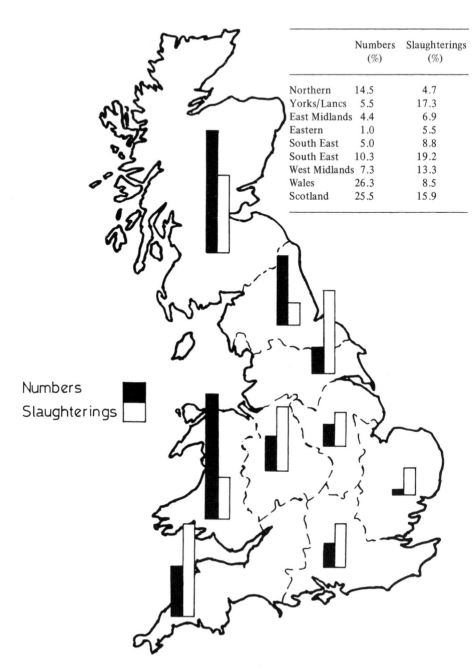

	Numbers (%)	Slaughterings (%)
Northern	14.5	4.7
Yorks/Lancs	5.5	17.3
East Midlands	4.4	6.9
Eastern	1.0	5.5
South East	5.0	8.8
South East	10.3	19.2
West Midlands	7.3	13.3
Wales	26.3	8.5
Scotland	25.5	15.9

Numbers

Slaughterings

Fig. 8.1 Comparison of the regional distribution of sheep numbers and slaughterings in the UK. (*Reproduced by permission of the Meat and Livestock Commission.*)

Table 8.4 Regional distribution of sheep, 1980.

	No. of sheep ('000)	No. of holdings	Average flock size
Eastern	315.3	1.3	238.3
South East	1 513.4	4.0	379.6
East Midlands	1 334.5	4.2	315.3
West Midlands	2 219.5	7.6	291.1
South East	3 109.8	10.9	285.6
Northern	4 395.0	9.7	453.8
Yorks/Lancs.	1 666.9	5.3	311.7
England total	14 554.5	43.1	337.7
Wales total	7 946.7	16.2	491.1
Scotland total	7 719.6	15.3	503.6
N. Ireland total	1 060.6	8.4	126.3
United Kingdom total	31 281.4	83.0	376.9

Points to remember

1. Factors affecting profitability.

   Sale price of product.
   Stocking rate.
   Control of perinatal deaths (good shepherding).
   Disease control.
   Flock depreciation.
   Sound nutrition at tupping, late pregnancy and during lactation.
   Economical use of labour.
   Sound management policies.

2. Sheep production systems.

(a) Mountain, moorland and arid areas.
   Sheep are usually the only economic agricultural exploiters of such land – otherwise, forestry or sport such as red deer, grouse etc. Poor living conditions, shortage of food, harsh climate, so sheep need to be small and very hardy.

(b) Grassland, mixed and arable farms.
   The farmer has a wide choice of options. He can conserve herbage

or hay or silage, he can grow crops for sheep feed or take advantage of arable by-products such as brassica crop residues. He can also return hoggets for feeding and is not forced on to a weak store market as the hill farmer often is.

9 Grazing systems for meat lamb production

The general approach to grazing

A farmer on either a grassland or a mixed farm who decides to go into sheep should give careful thought to the type of production appropriate to his farm and to the grazing system best fitted to give satisfactory results. He must ensure that the sheep enterprise is integrated into the rest of his farming and that his rotational system, method of establishing leys and timing of operations enhance the overall working of the farm and do not become a disruptive influence. Established farmers should not chop and change systems or breeds of livestock on the whim of the moment. On the other hand, they should not be deterred from making a periodic reappraisal of their situation.

What system of sheep farming is selected will depend on a number of factors. The farmer may have had previous experience of sheep and go straight into meat lamb production. On the other hand, he may be a stranger to these animals and wish to make a gradual approach, in which case agisted sheep such as hoggets for wintering may be taken. (Agisted animals are those taken on to a farm for feeding, usually on the basis of so much per head per week.) Alternatively, he may take wether lambs for fattening on clover aftermath and, by such means, familiarise himself with sheep and their handling before venturing into the more complicated business of lambing a flock. He may, however, decide to rely on the knowledge and expertise of a hired shepherd and go straight into meat lamb production. From the above it will be seen that there are a number of ways of entering the world of sheep farming. Notes on lowland sheep production, other than meat lamb, will be given in Chapter 12.

The pasture

There is a noticeable tendency amongst some graziers to take one or other of two opposite views of grass. On the one hand, some farmers become so enthusiastic about grassland and grassland societies that one wonders whether or not the aesthetic appreciation of beautiful emerald swards takes precedence over profitable meat production. On the other hand, some farmers treat grass when it grows as a gift from heaven to be consumed with no thought for its cost or long-term well-being. These extreme positions are both bad for profits.

Almost all grassland in temperate, well-watered areas of the world tends to revert first to scrub and then to forest if left ungrazed. Indeed, between World Wars I and II, this succession could be seen on the heavy clays of the English Midlands where undergrazed permanent pasture gave way first to brambles then to blackthorn, and oak saplings had appeared by the time World War II broke out.

The above is a gross example of change in the composition of a sward, but it also takes place in less obvious ways. There are various ways in which a sward may be altered other than by over-grazing.

PLANT FOOD

Most people are aware that legumes fix nitrogen from the atmosphere and that legumes such as clover in a sward provide nitrogen, not only for themselves but also for their companion grasses. On the other hand the modern practice of giving heavy dressings of nitrogenous fertilisers to grassland stimulates the grass to the detriment of the clovers. Heavy dressings of nitrogenous salts, with rapidly repeated defoliation, depress clover growth to the extent that legumes may virtually vanish. If plants are denied the opportunity to photosynthesise by too frequent defoliation, especially in the early spring, this will lead to a situation where the plants will be severely weakened if not actually killed. In this weak condition they cannot compete adequately with the rosette type of broad-leaved weeds such as shepherd's purse (*Capsella bursa-pastoris*) and the dandelion (*Taraxacum officinale*). This leads to rapid pasture deterioration.

The plant food nitrogen has been mentioned and, in addition to this, there are various other major elements which need to be added to the land to sustain production. Of these, potassium and phosphorus are the most important. On most soils liming materials to correct acidity are also required from time to time. A reminder that the application of nitrogen and potassium salts together in the spring-time tends to depress the uptake of magnesium by the herbage, with a consequent danger of hypomagnesaemia especially to lactating animals, must be given.

In the normal course of nature, mineral elements which tend to be leached from the top soil are replaced by plants with deep tap roots (provided the basic parent soil material contains these elements). The high mineral content of some broad-leaved plants, especially of the minor elements, is held by some to be important to the grazing animal. Experiments carried out to date have failed to support this view, but perhaps the experiments were too short in duration, or perhaps the shallow rooted grasses were receiving minor elements from impure fertilisers.

Besides shortages there are cases where problems can be caused by the excess uptake of minerals by plants. Typical of this is the uptake of molybdenum by plants where teart pastures are found (see also Chapter 2).

CLIMATE

In addition to the factors relating to plant growth, temperature, soil type and water supply will also affect grassland.

Most grass plants do not begin to grow until the temperature rises to about 5°C and Western European grasses achieve their best growth at about 13–18°C. This being so, the farmer cannot do a lot about grass growth until the normal spring temperature rise occurs. Not only must the air temperatures be right; there is also the question of the soil. Light, sandy soils warm up earlier in spring than do clays; a well-drained soil warms up earlier than a similar ill-drained soil, and artificial drainage can make a marked difference to the time of commencement of growth in the spring. In the northern hemisphere, south-facing land naturally warms up sooner than north-facing slopes. All these points must be borne in mind when planning a grazing programme.

Water

It is quite clear that water, or the lack of it, plays an important part in pasture growth. In areas such as the South East of England, where mid-summer droughts are not uncommon, one of the most useful insurances against failure is the availability of an irrigation system. Where the facility is available the soil moisture deficit should not be allowed to fall below 50 mm. (The soil moisture deficit is the depth of water (e.g. rain) required to raise the land to field capacity, i.e. the point where soil water moves under gravity.) It is quite wrong to wait until grass wilts and stops growing before applying water.

As to whether or not a sheep enterprise will justify an irrigation system is a matter for debate, but on a farm where equipment is in use for market garden crops, a dairy herd or some similar enterprise, advantage should be taken of its availability.

PASTURE PLANTS
Finally, in the establishment of grassland, the variation in performance of modern bred pasture grasses and legumes should be borne in mind. Earliness, habit of growth, time of maximum production and time of flowering, all these should be taken into consideration when sowing out new leys. Those who are not knowledgeable in establishing pastures should seek advice from neighbouring farmers, college extension officers, advisory services and similar agencies. It is to their advantage for sheep farmers to learn to grow grass really well.

Economic factors in meat lamb production

It is assumed that on the rich lowland pastures of the sheep producing world, the major product will be meat lambs. In areas where land is expensive the production will need to be intensive if it is to be economically successful. In addition to establishing a satisfactory pasture system, the farmer should try and ensure that he acquires a breed or cross of ewe suited to his system and technical capabilites. When making his choice the major factors of importance are the same as for all sheep enterprises but there are a number on which particular emphasis must be placed.

The first is the *fecundity and milking ability of the ewe*. It is obvious that if one starts with ewes that do not have the potential for large litters and the ability to feed the lambs produced, a successful intensive system is not possible.

The second is the *inherent growth potential and fleshing quality of the lambs*, there being no point in producing a large number of lambs of unsuitable growth rate and carcase conformation. Avoiding this usually means using a Down ram.

The third and fourth considerations, namely *disease control* and *flock depreciation*, are closely interrelated. The more ewes that are lost in the course of a year, the more replacements that are needed at the making up of the flock for the next season. Also, the more ewes that have to be discarded before the normal time for such reasons as mastitis, the higher the cost of maintaining the flock.

Intensive production also increases the dangers from internal parasites such as stomach worms and coccidia. Stress amongst ewes is also increased and physiological diseases such as hypomagnesaemia tend to increase compared with extensively run flocks. Lambs are also in greater danger from diseases such as joint ill, and mismothering can also prove troublesome.

This combination of factors means that what is being sought for intensive conditions is a sheep which is (a) very prolific and (b) highly productive in both quality and quantity of lamb meat, (c) has the ability to resist disease and (d) can live long. To meet all these needs requires a quite outstanding sheep.

Stratification has gone a long way towards providing such sheep; such crosses as the Scottish halfbred, Greyface, Mule and Welsh halfbred all exhibit these attributes to a substantial degree. Experimental work still continues in all sheepkeeping parts of the world, examples from Britain being the use of the Finnish landrace in improving fecundity and of the Texel (see Fig. 9.1) to improve fleshing quality.

THE CHOICE OF BREED OR CROSS OF SHEEP FOR MEAT LAMB PRODUCTION

The attributes of different breeds and crosses of sheep, their strengths and weaknesses, are difficult to assess. The views of experts are, at the best, opinions, there having been far too few critical and well-replicated experimental comparisons made between various breeds for anyone to speak with much authority as to which animal is best for a given situation. It seems fashionable nowadays to inveigh against the large number of breeds of sheep in Britain and, indeed, in the world. If, however, most breeds and crosses have as little worth as is often suggested, then economic forces should see to their removal and obviate the need for bureaucratic action in the name of agricultural efficiency.

In the meantime, the large variety of sheep breeds provide a good, deep gene pool where the enterprising breeder may fish with interest and, possibly, profit. The matter of breed assessment has been very ably discussed by Professor Spedding in *Sheep Production and Grazing Management* and students would do well to study his observations.

Fig. 9.1 Texel ram (*Courtesy of Mr J.A. Minto, Craigknowe, Biggar. Photo-graph: Douglas Low*).

Notwithstanding the above, some indication must be given of what sorts of sheep are likely to prove successful meat lamb producers. The first question that has to be settled is whether or not the farm will carry a self-replacing or bought-in flock. Flocks which are not self-replacing are usually referred to as flying flocks, although flying conveys a more ephemeral situation than should be used to describe a flock where ewes may be retained for up to seven or eight years.

If a decision is made to run a self-contained flock, one is limited to a small number of breeds from which to choose. This is true of Britain even though there are about fifty breeds in the country. Breeds such as the Clun Forest and the Kerry Hill possess all the requirements of meat lamb producers in fair measure; even these, however, produce a better meat lamb from a Down cross than they do when bred pure. The major objection to the pure bred flock is the problem of the ewe hogget. As previously stated, it is not norm-ally profitable to use the best land for stock rearing and, if it is done, the stocking rate needs to be high. This poses worm problems as the worst animals for fouling land with worm eggs are lambs and hoggets.

A lowland farm which possesses an area of heath or rough grazing land to which the hoggets could be sent could well prove profitable. In this case the ewe lambs to be used for breeding could be weaned and segregated before their contemporaries go off for slaughter.

The premier question regarding non-self-replacing flocks is the problem of the source of replacements. The normal practice with such flocks is to keep the ewes until the end of their useful lives appears to be in sight although many farmers tend to dispose of old ewes at too early a stage. A flock is normally kept up to strength by introducing gimmers on an annual or biennial basis. While not keeping the flock in exact age groups, it does ensure a reasonably uniform distribution of ages. Ewe hoggets are sometimes mated in their first year but this should only be done where it is possible to keep them separate from the main flock in the first year.

In selecting the ewes to be used, the farmer should consider *availability*: in other words, he needs to choose a reasonably popular cross where he can be assured of a satisfactory number of suitable replacement sheep every year. In a country such as Britain it will be a longwool mountain cross: in Australia and the Americas a longwool Merino or Merino derivative cross. Popular crosses in Britain are

> *Greyface* = Border Leicester × Scottish Blackface
> *Halfbred* = Border Leicester × Cheviot
> *Mule* = Bluefaced Leicester × Swaledale and
> *Welsh Halfbred* = Border Leicester × Welsh Mountain.

It should be noted that, in describing cross breeds, the name of the sire is given first.

Gimmers from the above crosses are available in large numbers and are offered in early autumn at sales in their particular areas. Unless one is buying regularly in large numbers, is very well-informed and in close touch with the area from which the sheep come, the best way of obtaining them is through a reputable dealer. Contrary to the received wisdom of some quarters, livestock dealers are no more dubious in their activities than other members of society. Indeed, the probity of the majority would provide a model for many traders. Once you and your dealer know what you really want he will endeavour to meet that requirement year by year, provided of course he is properly paid.

In a few cases meat lamb producers use cast mountain ewes direct, a ram such as a Dorset Down being used on Welsh Mountain ewes or a Texel on Scottish Blackfaces. The advantage of using such ewes direct is that the cost is lower than for crossbreds and they provide a meat lamb of excellent quality. On the debit side they are often difficult to contain, being able to get through very small holes in fences. They are also prone to making a real effort to escape and tend to be more susceptible to a number of health upsets such as enterotoxaemia and hypomagnesaemia than crossbred ewes.

Returning to cross breeds, the farmer needs to bear in mind how intensive his system is to be and whether or not he will be housing his sheep. Strange to relate, most of the mountain sheep crosses are quite amenable to a high stocking rate and to winter housing. This is particularly true of Mules and Halfbreds.

Finally, the mature size of the ewe is of importance. The larger the sheep the greater the cost of food and the fewer that can be kept per unit area. Having made this point, it does not seem reasonable to employ a very large ewe to mother two lambs that will be slaughtered at 35 kg each live weight. On the other hand, there is a limit to the emphasis that should be put on small size. In some cases such as in the production of very early lambs, the larger ewe with more milk will get off a larger lamb more quickly than will the smaller ewe. A large ram such as an Oxford Down can also be used to good effect. This will ensure that a high price market will be captured.

The selection of rams for meat lamb production

Having decided on the breed or cross of the ewes to be used, a decision on the ram must next be made. Which breed of ram to select depends mainly on the anticipated fate of the lamb but in almost every case it will be a Down breed. Whilst very early meat lambs may be sired by an Oxford, Hampshire, or Dorset Down and lambs from small ewes by a Southdown, the majority of meat lambs in Britain are sired by a Suffolk. The reason for the popularity of the Suffolk is that while lambs can be carried to quite substantial weights without becoming over-fat, relatively light carcases can be produced from this sire without their being too lean.

A simple computation devised by the Meat and Livestock Commission gives the approximate appropriate slaughter weight of lambs from various crosses which is as follows:

The slaughter weight of lambs grown without check and with no store period is about one quarter the combined breed weight of the parents.

This gives for the Southdown X Welsh Mountain:

$$\frac{61 \text{ kg} + 45 \text{ kg}}{4} = \frac{106}{4} = 26.5 \text{ kg live weight (i.e. at 50\% killing out weight, a 13 kg carcase)}$$

Similarly, with the Suffolk X Mule:

$$\frac{91 \text{ kg} + 85 \text{ kg}}{4} = \frac{176}{4} = 44 \text{ kg live weight or 22 kg carcase weight}$$

At slaughter, the male lambs will be somewhat heavier and the ewes somewhat lighter than these average weights.

The second most popular ram for fat lamb production in Britain is the smaller, rather more quick-fattening Dorset Down.

With the advent of a marked preference for a really lean lamb by the consumer, the Texel has come very much to the fore. The Texel gives a leaner carcase than does a lamb from a Down cross, the killing out percentage is also better, but it takes a Texel cross longer to reach slaughter weight.

If the lambs are to be sold on the hoof careful consideration must be given to local market preference. If, on the other hand, the lambs are to be sold on the hook the farmer can experiment to find the breed of ram that suits him best. In comparisons of this type a reasonable number of rams must be compared over a period of time in order that the idiosyncrasies in the breeding performance of individual rams can be eliminated.

Individual grazing systems

All systems of pasture management have to take account of the seasonality of pasture growth which, in turn, depends on the climate. In maritime countries the pattern generally shows a build-up in the rate of growth through the spring which reaches a climax in early summer and then declines. This summer decline is followed in late summer and early autumn by another resurgence of growth which dies away in late autumn and early winter (see Fig. 9.2). As has been emphasised by Professor M. McG. Cooper, this production pattern is more nearly coincidental with the food requirements of

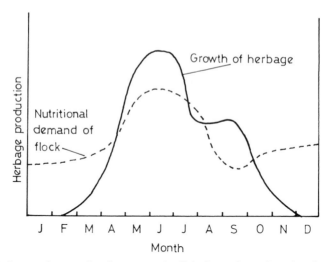

Fig. 9.2 Seasonal growth of pasture in Britain at low elevation in a normal season related to the nutritional demands of a meat lamb producing flock. The lambs are all sold as they reach slaughter weight.

a flock of ewes and lambs than for any other livestock.

The objective should be to lamb the ewes at such time that the greatest pasture demand of the ewes and lambs coincides with maximum pasture growth. The main problem the situation presents is that the stock are unable to utilise all the grass at peak growth. The solution to this problem of early summer surplus is the most important in good pasture management and this is usually achieved by increasing the stocking through the introduction of cattle or by setting aside a portion of the grazing for conservation.

When the peak and its problems occur varies in different places. In Britain, the southern and western areas experience periods of greatest growth earlier than the north and east; lambing in the former areas is therefore, on average, earlier than the latter but these times are modified by elevation. Figure 9.3 indicates the type of growth curve found on hill areas. The herbage growth curve in these high areas is also influenced by aspect; some steep north facing slopes have very curtailed growing seasons.

In planning a system it is important to know what other stock, if any, has to be grazed as on farms which carry a mixed stock such as beef and/or dairy cattle their various requirements must also be taken into consideration. The classical fattening pastures of Leicestershire and Northamptonshire, for example, were manipulated and

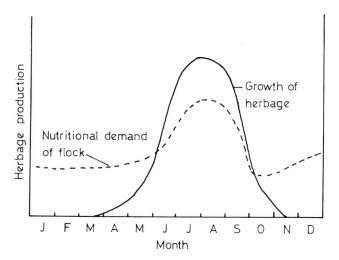

Fig. 9.3 Seasonal growth of pasture in Britain at high elevation in a normal season related to the nutritional demands of a self-replacing hill flock where lambs are mainly weaned together. (*Note*: The graphs in this and the preceding Fig. 9.2 apply to normal seasons but severe early or late frosts, or a protracted drought, can cause serious distortions to these patterns.)

kept in good order by the adding to and subtracting from the mixed stock carried by various fields.

The extremes of the different grazing systems are *set stocking* and *creep grazing*. There is no doubt but that the *best individual lambs come from set stocking*, while the *biggest output of lamb meat per hectare comes from creep grazing*. One reason for the reluctance on the part of many farmers to accept creep grazing is that they are more used to making visual appraisals than to making physical quantitative checks. The result is they are too often difficult to persuade that the best-looking animals are not necessarily the most profitable.

SET STOCKING

In set stocking at its simplest, the farmer or the shepherd turns out to the field the number of ewes and lambs he considers appropriate and leaves them there. Lambs will be removed as they are ready for slaughter and, perhaps, the ewes for drying off. This is the basic traditional system in Britain.

As with all grazing systems, the crucial question is *how many and for how long*. On an understocked pasture the grass gets away from the animals and this is to the detriment of the lambs which like reasonably short, fresh growth. On farms where the grass forms part of a rotation, the ewes and lambs should go on to a maiden ley as this gives them a good start by being free of worm parasites. If the stocking rate is too heavy or the sheep have been put on too early, the proper establishment of the ley will be impaired, if not prevented. If the sheep are on a farm that consists of permanent, or mainly permanent, grass the flock should go on to ground that has carried no sheep during the previous year. This will ensure that no problem arises from nematodirus infection.

In terms of the number of sheep stocked per hectare, the majority of farmers tend to understock keeping about 6–9 ewes with their lambs per hectare. The more able farmers achieve 12–13 ewes per hectare and this is normally reflected in the increased profitability of those with the higher stocking rate.

On large farms with a big flock the ewes and lambs will be divided up, and this should be done on the basis of the greatest need to the best feed. Ewes with triplets and twins should go to the best land, gimmers with single lambs to the next best and ewes with singles to the poorest. Where the production is aimed more at store lambs, the stocking rate can go up to 14 ewes with their lambs per hectare.

In many areas of mixed farming where sheep are important, the surplus grass at peak production is controlled by the introduction of store cattle or dairy heifers. Where cattle are not available the mowing machine or silage harvester has to be used.

Rotational grazing
The difficulties which arise from the uneven growth rate of pastures and the necessity for the removal of peak growth or the supplementary feeding when the peak has passed, have caused many farmers to modify their set stocking. Where there are a number of fields available or where big fields can readily be divided up into paddocks, they control herbage use by rotating the sheep from paddock to paddock. This approach causes problems with regard to investment in fencing and water supply. Where there are only sheep, fencing can be provided by ordinary sheep netting on wood or metal stakes or by electrified sheep netting. Where there are cattle at least one strand of barbed wire is required and this necessitates strainers, making for an expensive fence. Water can be laid on by using a polypropylene pipe laid in a plough furrow to a series of troughs, or it can be provided by a mobile tank.

The chief problem with the paddock grazing system is that it leads towards a feast or a famine situation, and the ewes and lambs are in direct competition with each other for most of the time. Some farmers do, however, make the system work quite well. First, they ensure that there are ample paddocks relative to the size of the flock or flocks. Secondly, the majority have a number of cattle that will be finished off in yards during the winter; these are moved round the paddocks as appropriate. In addition to the control exercised by the cattle, some of the land is taken for conservation. The best form of conservation relative to pasture management is silage or dried grass because cuts for silage and dried grass can be taken early in the season relative to hay. This ensures regeneration of the aftermath in time for it to be used by the ewes and lambs. Silage and dried grass, on the other hand, is taken when plant growth is still vigorous and the weather and soil moisture normally conducive to quick recovery and a good response to fertiliser nitrogen. Hay, however, has to be taken too late in the season for this to prove really effective.

Stocking rates of 15–17 ewes to the hectare can readily be achieved.

CREEP GRAZING

In an ideal situation lambs would have access to short, freshly grown herbage from which the ewes were excluded. The exclusion of the ewes would ensure that the pasture was virtually uncontaminated by intestinal and stomach worms. Sideways creep grazing (see Fig. 9.3) was evolved with this ideal situation in mind.

Sideways creep grazing

A *creep* is merely a device whereby small animals such as lambs and calves can be allowed away from their dams either to graze or to have access to concentrated food or hay.

Creeps for lambs to use in creep grazing can be simply holes made in the fence through which the lambs can pass but the ewes cannot. In more sophisticated systems creeps are provided by using factory-made metal hurdles. The lower section of the hurdle has upright spars, preferably rounded, which are movable so that they can be spaced further apart as the lamb grows in size.

The first problem to be faced in devising such a system is that there is a limit to the distance a lamb will move away from its mother. In other words there is a limit to the size of paddock that can be used. The shape of the area is also important; for instance,

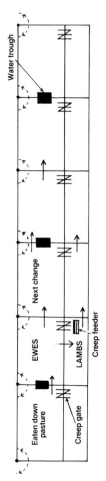

Fig. 9.4 Sideways creep grazing paddocks.

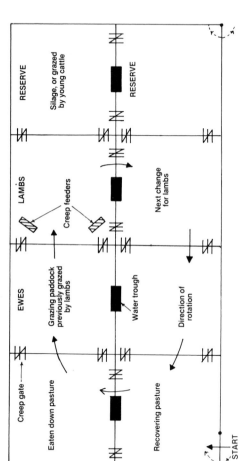

Fig. 9.5 Forward creep grazing paddocks.

a large oblong with a creep fence at one end would not be at all suitable. If, on the other hand, the area were divided longitudinally, access of lambs to the creep area would be much enhanced.

The main difficulty with sideways creep grazing (see Fig. 9.4) is the question of area; as the lambs grow they need more grass and it is difficult to make an arrangement whereby this is easily done. It must be achieved either by extending the area to which the lambs have access or by the use of other stock such as cattle. The system is useful where the production of lambs with a minimum burden of internal parasites is required for experimental purposes. It does not, of course, ensure total freedom from parasites as there are no means of preventing the lambs from grazing beside their mothers.

Forward creep grazing

The forward creep grazing system has been evolved to ensure adequate grazing for heavily stocked ewes and lambs and to eliminate competition when the lambs' requirement for grass is increasing rapidly. It depends on a series of paddocks which are grazed in rotation, the lambs being allowed to press forward through creeps to the new grass from which the ewes are excluded.

The minimum number of paddocks on which the system can be run is six, and where accommodation grass is not available in case of mishap the number should be eight. Fig. 9.5 shows the preferred layout of the system but it could, of course, be modified to fit in with the shapes of existing fields. These paddocks can be temporary or have a certain degree of permanence.

The smallest paddock that can be used conveniently is one of about 0.4 hectares (or one acre). The largest area the author has seen operating well is 1.6 hectares per paddock in a six paddock layout. This enterprise carried two hundred ewes and their lambs. It is obvious that the numbers of sheep in a system cannot be multiplied up indefinitely. If the paddocks become too big, a situation will arise whereby the paddocks are so large that many lambs will scarcely see the creep gates, much less go through them.

In a system of small paddocks, the creeps should form the gates between the paddocks as lambs tend to work along fences and are thus directed to the next paddock. In large paddocks there should be a creep alongside the inner fence, as shown in the diagram. Water troughs should be provided, one serving two paddocks. These should lie longitudinally along the centre fence, fed by a polythene pipe buried beneath a plough furrow. The creep should be of vertical rollers incorporated in the gates. These give little trouble and, while the initial cost may seem high, they will last for twenty years or more. The only problem arises when the ewes are shorn as a small ewe with its fleece removed is very similar in size to a large lamb with its coat on. The result is that some ewes creep forward. This can usually be cured by tying a wooden batten across the creep some 40 cm from the ground. The ewes do not like getting down on their knees to pass through and are usually discouraged.

A creep feed for concentrates should also be provided. It should be of the type with a veranda to keep the concentrate from getting wet and to help prevent puddling in the area round the creep. The creep should be placed within the lambs' grazing area near the creep gate but moved a short distance daily to prevent poaching. The feed provided should be of three parts rolled barley to one part flaked maize, the flaked maize being reduced progressively until it is eliminated — unless, of course, it becomes cheaper than barley. The only other worthwhile additive is 3% of a simple commercial mineral mixture providing calcium, phosphorus and common salt. The addition of protein is often recommended, but is often unnecessary. Lambs on good, young grass and ewes' milk are not usually limited in growth by protein shortage but by energy. Failing barley, oats or dried sugar beet pulp can be used.

Turning to the composition of the ley, many types of seed mixtures can be used for intensive grazing. Ones based on perennial ryegrass, timothy, meadowfescue, or even the old fashioned Cockle Park mixture, are all satisfactory provided they are properly handled. In most parts of Britain, experimental evidence points to perennial ryegrass being the most productive but observations on grazing animals suggest that a timothy, meadowfescue pasture is more acceptable to sheep when they are allowed a free choice.

Operating the system

The object of the forward creep grazing system is, as we have said, to keep a satisfactory quantity of good grass available to both ewes and lambs throughout the change in their relative requirements. In the early stages it is the food of the ewe that is of major importance so once the paddock is reasonably bare the sheep must be moved. If the nutrition of the ewe falls off, the milk yield will decline rapidly and the growth rate of the lambs will fall, as in early life adequate milk is the important factor for lamb growth.

The milk yield of the ewe normally peaks at 2–3 weeks and begins to fall off at 5–6 weeks. Lambs in the early stages of their lives, say up to the age of about 12 weeks, will grow at about 0.35–0.50 kg a day. Later growth falls to 0.25 kg a day, and after weaning it will be even less. This means that the lambs must be very well fed in early life. Trying to meet these requirements mainly from grass means keeping the sheep on a particular paddock for 3–5 days, depending on conditions.

Stocking rates and timing
The time of starting the ewes on forward creep grazing is important. There should be adequate grass available otherwise the moves will have to be made too rapidly, with the consequent weakening of the grass plants. On the other hand, as previously stated, if the sheep arrive too late the herbage will have grown away from the lambs and become too coarse.

The second feature of the timing is the age of the lambs; the lambs must not be too old when introduced to the paddocks. An age of 2–3 weeks is optimal as older lambs aged 4–6 weeks have become conditioned to grazing beside their mothers and are reluctant to move forward through the creeps.

As the season progresses and the bulk of lambs get to be over 6 or 7 weeks or age, their need for grass increases rapidly so it is at this stage that the lambs need large quantities of short, leafy grass. The ewes, on the other hand, have lower nutritional requirements as their milk yields are falling. The ewes are therefore made to work harder and bare down the pastures, care being taken that the lambs are never short of grass. This grass must be highly digestible and we should remind ourselves of what was said in Chapter 2, namely that as the digestibility of a food falls, the rate of passage through the gut slows down and the total intake of nutrient falls. The automatic result of this fall in intake is a slowing down in growth rate. This fall in growth rate is serious for two reasons. One is that the faster the growth rate of any particular animal the higher the proportion of its food devoted to production purposes. It might be argued that in all probability there would be enough grass in the long term to fatten the lambs successfully. This could well be so, but in a country like Britain there comes a time in the summer when the value of the lamb carcase drops quite markedly due to flooded markets. Therefore, the sooner a lamb is fit for slaughter the more profitable it will be. The second point is that in areas subject to summer drought, the grass may run out before the lambs get to slaughter weight. This results in the farmer being left with a number of store lambs which are nothing but an ambarrassment.

It will be seen from the above that the time spent by the animals in each paddock before being moved forward will vary according to grass growth and animal demand. The time normally varies between 3 and 5 days.

When considering stocking rates the make-up of the flock must be borne in mind. There is no splitting of twins and singles. This mixing of singles, twins and triplets ensures that as the season advances and the lambs become bigger there is some reduction in stocking rate. The single lambs are the first ones ready for slaughter and they and their mothers are removed from the flock. The twin lambs are next and will be ready shortly after the singles. The average good quality ewe, properly fed, will give more milk when she is suckling twins than when she is suckling a single, but one with triplets rarely gives enough milk for the latter to grow as fast as twins. This is another reason which justifies the use of creep fed concentrates as it ensures that all lambs are sold in a reasonable time.

When the pasture is ready for stocking, the rate depends on soil and climate. Over a period of years one learns when the grass may be grazed with little fear of such adverse conditions that it will be all eaten off and the system collapse. The number of ewes and lambs that can be kept will depend on the quality of the grassland. For a forward creep grazing system one normally thinks in terms of a ewe flock with 175–200% lambs at foot. The paddocks should be able to carry 20–25 ewes and their lambs to the hectare. The author has managed 30 ewes/hectare for a season but it tends to be a somewhat tightrope exercise, not to be recommended.

Earlier in the chapter, the use of eight paddocks rather than six was mooted as an insurance. Failing this, an area of permanent pasture or silage aftermath is almost essential. This can be used under extreme weather conditions for the whole flock or, in any case, for weaned ewes. In the author's twenty years' experience in South East England, it was necessary to remove the flock from a six paddock system on one occasion only during the grazing season. The sheep were removed for four or five days due to exceedingly heavy and persistent rain turning clay land paddocks into a quagmire.

The reader may express some surprise that no reference has been made to worm parasites until now. Parasitic worms do not usually cause a lot of trouble under this system. The ewes should receive a worm drench before tupping and again before lambing and, in most years, this should suffice. Some lambs may need drenching at about 8 weeks of age and these are usually twins and triplets. Rates of infestation depend very much on weather conditions and the condition of the lambs should be the guide. Unnecessary routine drenching is counter-productive.

The reason why worms do not constitute a serious problem is not directly because of the rotational aspect of the grazing but because of the creep system. The system, when properly run, keeps a good supply of succulent grass of a suitable length in front of the lambs and this, together with some concentrate feed, ensures a satisfactory nutritional status. As underlined in Chapter 2, the well-fed animal is much less susceptible to parasites than the under-fed one. Also it is probable that a plentiful supply of grass of about 10 cm in length renders it unnecessary for the lamb to graze close to the ground and, hence, it picks up fewer larvae than does one on short commons.

The view has been held by a number of persons that rotational grazing has a beneficial effect in controlling internal worm parasites, in that the eggs passing out in the droppings of the ewes would hatch out and the larvae die off before the sheep and lambs returned to the original paddock. The trouble with this theory is that the hatching time of most worm eggs depends very much on temperature and moisture, as also does the survival of the larvae. It is not difficult to envisage a weather situation whereby hatching is delayed, then conditions turn suddenly favourable to the parasites with the result that the sheep flock return to an eruption of larvae much greater than they would have met in set stocking.

Health hazards of the forward creep grazing system
Discussion of parasites turns attention to other health hazards that can cause problems in the system. It has to be accepted from the onset that forward creep grazing is an intensive system where the animals are placed under greater stress than are animals under more extensive regimes.

There is the stress caused by crowding and, in this context, it is important to see that the bonds between ewe and lamb are firmly established as soon as possible. In other words, the lambs must be well mothered up before they are introduced to the system or the ewes will be using good grazing time in seeking their offspring rather than eating. Also in connection with stress, ewes will be more susceptible to physiological upsets such as hypomagnesaemia and hypocalcaemia. Calcined magnesite should be fed in the concentrate ration for a week before lambing, until six to eight weeks after turn-out. This concentrate feeding has an added advantage in that the lambs at foot learn to eat concentrates which makes them more avid for the concentrate creep feed. This gives an opportunity to make the point that if animals are expected to eat concentrates in later life they should be taught to do so when young. The difficulties of feeding concentrates to hill ewes in times of stress or getting wether lambs to accept turnips result from their not being familiar with the food and this will be returned to later.

Pulpy kidney disease in the lambs and enterotoxaemia in the ewes will definitely prove troublesome unless protection is given. The ewes and lambs should be vaccinated with one of the modern mul-tiple vaccines which also gives protection against the other clostridial diseases such as tetanus and lamb dysentery. Precautions against blow fly strike must also be taken.

Pasture management
Little need be said about this aspect of the forward creep grazing system as most has been implied in the animal management. One point that has to be underlined, however, is fertiliser treatment. It has been noised abroad by opponents of the system that vast quan-tities of nitrogen are required. This is not so. The following repre-sents a typical season's treatment of the plots with nitrogenous fertiliser.

Applications of potassic and phosphatic fertilisers are made in the autumn in accordance with requirements indicated by soil analysis. A rough figure would be 50 kg/hectare each of P_2O_5 and of K_2O.

The application of nitrogen in a typical spring was as follows:

30th March	50 units per acre
10th May	50 ,, ,, ,,
4th June	50 ,, ,, ,,
30th June	50 ,, ,, ,,
27th July	50 ,, ,, ,,

Total 250 units per acre

These dates are for a farm in mid-Essex. Farmers in the South West of England would start applying fertilisers much earlier. Starting dates for fertiliser application need to be related to climate and soil type.

The most spectacular feature of properly conducted paddock grazing is the rapid recovery of the sward. There is a tendency for a spiralling effect to occur, with the growth getting better and better until the season change tends towards flowering in the grasses. On the other hand, a pasture improperly stocked and handled soon goes into a decline in production and that decline can be equally spectacular.

KEY POINTS FOR SUCCESS IN FORWARD CREEP GRAZING
To sum up, the main factors which will determine success in forward creep grazing are outlined below.

1. Introduce a suitably heavy stocking rate at the right time

A suitably heavy stocking rate must be introduced at the right time. This requires a flock of ewes of high fecundity all lambing as near together as possible. To this end a lot of farmers use vasectomised rams which are put out before the entire tups are introduced. This practice does not add to the number of lambs produced but it does seem to concentrate lambing. Another possibility is the inhibition of oestrus over a period in order to bring in all the ewes together. This is achieved by the use of a synthetic substance such as cronolone which is introduced into the vagina in pessary form. The pessary is inserted with an applicator and a string is attached to the pessary to facilitate withdrawal. The pessary is left in position for 17 days during which time the cronolone is absorbed through the vaginal wall inhibiting oestrus and ovulation. When the pessary is removed the ewes come into heat and ovulate within 2 to 3 days. This practice has its problems, such as the increased number of rams needed, but this can be overcome by using vasectomised rams to detect heat and then hand mating or inseminating the ewes artificially. A successful operation of the controlled mating system not only has the advantage of having all the lambs of similar ages but the concentration of the lambing also makes for better labour utilisation. It also means that a more rational approach to the pre-lambing 'steaming up' of the ewes can be adopted. The overall growth of the lambs will also be more uniform.

2. Ensure forward movement at the appropriate time

Having ensured a suitable stocking rate and introduced the flock at the optimum time, the operator must keep a careful watch on the flock to ensure forward movement at the appropriate time.

3. Control of fly strike and foot rot

The farmer must also be alert to problems which are common to sheep under any system. Amongst the most important are foot rot and blow fly strike. The ewes and lambs should be sprayed against fly strike a week or so before the first attacks are expected. The attacks normally occur in warm, humid weather. If there has been any scouring, badly soiled wool should be clipped from the breech area. Driving the flock to a building for such purposes as weighing, foot dressing and spraying, should be avoided if at all possible. The secret of success with milking ewes and fattening lambs is to leave them in peace as far as is possible.

4. Leave the animals in peace: avoid unnecessary stress

A portable footbath in association with portable hurdles should be used to control foot rot. These hurdles are also necessary to form a holding pen and race for weighing the lambs. No meat lamb producer should be without a portable weighing machine which can be set up in the paddocks.

5. Weigh and point score for condition every week

Having made the point about disturbance, weighing is unavoidable. As soon as the lambs get near finished weight they should be weighed and point scored for condition every week. In addition to taking out those which are ready for the butcher, those which are anticipated will be ready the following week should be marked with a paint stick. Marking in this manner means the farmer knows approximately how many lambs he can supply for slaughter the following week. If the above procedures are followed, the flockmaster will not be embarrassed with over-weight or over-fat lambs. It is assumed that the lambs will be sold 'on the hook' i.e. by dead weight; this method, when averaged out, usually gives the best return. Regular weighing and handling will ensure that the slaughterhouse receives a uniform product on which it can rely and that the farmer builds up a firm market. This weighing and disposal procedure applies throughout the season until a handful of the worst doers are left. These should be consigned to the abattoir for what they will fetch. It is quite uneconomical to struggle on in an attempt to get the last individual away finished.

The normal duration of this intensive grazing is three to four months; in Britain most low ground lambs should be sold by the end of July and, in the South of England, by mid-July.

Having summarised the points of importance in the good management of creep grazing, it remains to be said that, as with other livestock husbandry enterprises, the main factor contributing to success is that the person in charge wants it to succeed. Unless the operator is fully committed and prepared to use the necessary discipline the system will not succeed.

Experience with creep grazing in Britain would suggest that it can show a gross margin that can compete with other forms of animal production even on relatively expensive land.

Professor Cooper's suggestion that: 'It would be possible for a farmer with 120 acres of land in grass to carry 400 breeding ewes plus their lambs on half the grass during the spring and summer with the balance of the pasture being used for conservation and for such cattle as the farm might carry' is quite feasible.

The spring and summer creep grazing of the flock having ended, the ewes are put on to any poor land available. If the same creep grazing area is to be used for lambs the following year it should be kept completely clear of sheep until the spring. The practice of using the same area in successive years must not be followed if there is any suspicion of nematodirus being present.

The best place to accommodate the ewes in late summer and autumn, prior to flushing them, is old grassland due to be ploughed out. This takes us to what happens to the sheep in winter or, perhaps what is more important, what happens to the land in winter.

On free-draining land with well established leys this is usually no problem but on heavy land, such as some boulder clays or where all the grass is short rotation leys, the problem of winter damage can be acute. This raises the question as to whether or not the ewes should be housed for the winter. This housing of ewes is the subject of the next chapter.

Points to remember

1. Grassland management:

   Grass is a *crop* and must be *farmed*.
   Avoid over-grazing, especially in early spring.
   Avoid under-grazing, especially in summer or the pasture will deteriorate.

Apply adequate fertiliser but beware of excessive dressings of nitrogen which can over-stimulate grass to the detriment of clover.

Fertiliser applications must be at appropriate times and quantities.

Irrigation, where available, but must be started *before* a big soil moisture deficit begins.

2. Economic factors of meat lamb production:

High fecundity and milking ability.
Inherent growth potential and fleshing quality of lambs.
Good disease control.
Low flock depreciation.

3. Choice of breed or cross for meat lamb production:

Must be appropriate for market.
Sires must be appropriate for market objective and size of ewes on which used.

4. Growth of grass:

Growth at low levels starts earlier and lasts longer than at higher levels but normally shows a mid-summer depression.

At high levels, growth starts late but peaks rapidly in summer, falling off earlier in autumn. Does not normally show a mid-season depression.

5. Grazing systems.

There are two main systems:

Set stocking tends to give best lambs but also the lower output of meat per hectare.
Rotational grazing gives a higher output per hectare and is normally more profitable.

In general, aim to keep adequate leafy grass always available for the animals, and favour the lambs instead of the ewes as the season advances.

6. Rotational grazing systems.

In rotational grazing, sheep are moved from paddock to paddock. Creep grazing systems are refinements of rotational grazing which allow the lambs to graze separately. The best known are sideways creep grazing and forward creep grazing.

Sideways creep grazing gives good control of parasites but is economically difficult to justify and now almost obsolete.

Forward creep grazing gives good control of parasites and pasture but is expensive in equipment (water troughs, fencing etc). For success, it is essential to start at the right time, with the right numbers of ewes and lambs, and with a paddock in reserve.

10 The management of housed sheep

The reasons why sheep have been, and are, housed are various. In areas where there are predators such as wolves or coyotes it is essential to bring the animals into a protected area for the night or to have a large number of persons guarding them. In areas where sheep are milked daily the practice is, as with dairy cattle, eased by holding the flock overnight. In countries such as Scandinavia and North America where heavy winter snow is the order of the day, sheep housing is not an option but a necessity.

In the sheep-keeping countries of the southern hemisphere, however, sheep are regarded as totally outdoor animals. In Britain, a similar attitude has persisted for a long time. There have, of course, been exceptions, such as the housing of ewe hoggets in the Yorkshire Dales. In the last two or three decades there has been a revival of interest in housing especially for intensive ewes. One of the causes of the interest has been the re-introduction of grass breaks on arable farms. In the past the dairy herd was by far the most efficient utiliser of the grass ley, but the high capital costs and the problem of skilled labour in the establishment of a dairy herd makes this enterprise less and less attractive nowadays. With beef cattle the capital costs tend to be higher than for sheep, fencing and water supply being also more expensive. On the other hand the gross margins from sheep stocked on the conventional extensive system are much too low to justify a sheep enterprise on expensive low ground — so, if sheep are to be used the system must be intensive, hence the desirability of housing.

Factors on which the decision to house must be based

THE PASTURE

The first and most marked advantage to be gained from housing is the protection of the ley. Leys rested over the winter are not poached by hooves nor rutted by tractors carting food. The result

Fig. 10.1 Inside a sheep house, Faccombe Estates, Hampshire. (*Courtesy of Meat and Livestock Commission.*)

of resting is that they come into production much earlier than do leys which have been grazed. Also their total production for the grazing season is greater. It can be argued that the sheep could be overwintered on the ley due for ploughing out but this has two disadvantages: one is that it is normally much more profitable to plough out in the autumn and grow winter wheat than to grow a spring cereal. The second is that as the sheep will need supplementary feeding and if bulky material such as silage or roots are carted to them the areas round the gates and feeding troughs can become like a quagmire to the detriment of the sheep.

From the stand point of the grassland there can be no argument but that removing the sheep over winter is beneficial. There are other aspects of the question of whether or not to house which are much more debatable and we will now look at them.

NUTRITION
The protagonists of outdoor wintering always raise the question of food costs. The quantity of hay eaten over a 90-day period is about

150 kg per ewe, but the outdoor ewe on an intensive system will probably require 75 kg over the winter. One point worth noting is that the wastage of hay in outdoor feeding is substantially greater than with indoor feeding. The quantity of concentrates used will be the same for both systems.

Prophecies of heavy losses of housed ewes due to pregnancy toxaemia brought about by lack of exercise have not come to pass. In a well-regulated flock the incidence of this disease is less than for a comparable outdoor one, but it can happen. If there has been a mild back-end the ewes may come in too fat to start with. If such ewes are steamed up too heavily trouble will surely ensue.

On the question of exercise, the indoor ewes get ample provided the floor space is reasonable. If anyone cares to identify half a dozen ewes and watch them for a few hours, keeping a plot of their changing positions, their proclivity for taking exercise becomes apparent. Indeed, indoor ewes will take a lot more exercise than will ewes up to their hocks in mud on a turnip break.

LABOUR

The amount of labour required for feeding and general shepherding of an indoor flock is less than for an outdoor one, always provided feed such as hay is stored to hand or, where silage is used, a self-feed facility is incorporated in the building. The care which can be devoted to ewes at lambing is greater and additional assistance is more readily available to the shepherd. The lambs and the shepherd are both protected from the elements, which makes for a better survival rate amongst the lambs and a more comfortable and contented shepherd.

HEALTH

The prevention of perinatal deaths can be a very important advantage of housing in intensive production systems but good hygiene is essential. The health of housed ewes presents no great problems provided the ventilation is satisfactory, that the sheep lie dry and are not overcrowded in respect of floor area and trough space. This is, of course, provided that there is no outbreak of infectious disease such as contagious abortion or *metritis* (inflammation of the uterus). Infectious diseases naturally spread more rapidly indoors than out. The lambs derive great benefit from being born indoors and this is especially true of small lambs from triplets. These can be placed under an infra-red lamp and have colostrum administered by stomach tube much more readily than can be done outdoors. Housing also avoids the serious damage that can be done to young lambs in the first 48 hours of life by cold rain, sleet and hail.

The greatest danger with housed lambs comes from *septicaemia*; joint-ill can become a major problem unless hygiene is of the best. Similarly, *coliform scouring* can cause havoc if it once gets a hold, especially where lambs are reared intensively indoors throughout their lives.

Pasteurella infection which produces a type of pneumonia has caused trouble with some housed animals and it is a difficult condition to handle. The causative organism, *Pasteurella haemolytica*, is present in many healthy sheep and although the early vaccines against the disease were not very effective, a more recent vaccine is showing good promise. In view of the difficulties that this condition presents, any suspicion that it has become active in a group of sheep should prompt immediate veterinary investigation. The symptoms of the disease are the usual high temperatures, high respiration rates, coughing and watery eyes. As with many sheep diseases, the onset may be precipitated by stress. An airy, dry environment is the best protection against the disease.

Orf has occasionally caused trouble in housed lambs in Britain in recent years. It is a virus disease which causes eruptions on the lips and around the coronet and its scientific name, *contagious pustular dermatitis*, is an all-embracing description. A fair degree of control can be obtained through the use of a vaccine but veterinary advice should be sought because it is a live vaccine; the action of live vaccines require more careful monitoring than dead vaccines.

This question of hygiene must be taken very seriously as there is no point in saving lambs at birth by protecting them from the elements to lose them a few days later from infection.

It should be noted that the number of lambs born and their birth weights will show little difference between outdoor and indoor wintering, always provided the nutrition is comparable.

CAPITAL COSTS OF HOUSING

What has been said for and against housing may prompt the beginner to feel that housing should be the preferred way of wintering sheep. The problem, however, is that housing can be very expensive and the profits in most countries from sheep keeping are not vast. Where the capital is available it is only too easy to embark on a housing scheme which will saddle each ewe with a debt that is unpayable. In order that compromises can be explored we must look at the minimum requirements of a sheep house.

The basic requirements of a sheep house

VENTILATION
The paramount requirement is that ventilation must be very good but there must be no draughts at floor level. This means that the air intakes must be above 1.5 metres in height and with open-sided buildings there is a baffle wall of at least this height. Ventilation must be such that in all parts of the building there must be air movement above the sheep's head. Dead areas such as are found against the inside walls of lean-tos must be avoided.

FLOOR SPACE
There must be adequate floor space but where slatted floors are used this must not be excessive otherwise the dung will not be trodden through between the slats into the dung pit.

The area allocated per sheep will, of course, depend on the size of the individual sheep and the number kept in each compartment. Some people do not like to go beyond 30 ewes per pen but others have had no difficulties with numbers up to 60. The breed, cross, and strain of sheep being kept is relevant as some sheep are much more amenable to close confinement than are others. Suggested area allocations per sheep are 1.2 m²–1.5 m² for in-lamb ewes on solid floors, with 0.8 m²–1.0 m² for those on slatted floors; areas for ewe-hoggets and similar sheep are proportionally smaller.

LIGHT
The provision of adequate light is not usually a problem under British conditions where the sheep are normally housed for only three months. In higher latitudes and under a longer indoor regimen, Vitamin D deficiency could occur. Over the three month period, no problems are likely provided the concentrate diet is fortified with synthetic Vitamin D.

Where housing is for a longer period with a very short winter day, Vitamin D injections may be advantageous. Provided the sun is able to shine directly on to the sheep (without passing through ordinary glass) for a period of the day and the light intensity is such that reading normal print in the middle of the day is possible, lighting may be assumed to be adequate.

TROUGH SPACE

In addition to getting the area right, the shape of the pens must be such that troughs and racks can be fitted in a manner so as to provide sufficient feeding space for each ewe. The positioning of the troughs must also have reference to ease of filling.

While combination hay and concentrate racks (see Fig. 10.2) placed alongside a central gangway make for easy foddering, they can give rise to the problem of old or boss ewes lying alongside the racks and preventing the younger and more timid ewes from getting to the hay. There is therefore an advantage in keeping hay racks away from the walls.

The combination rack which is, in effect, a hopper placed over a trough with a vertical gap between and also a gap at the bottom of the hopper, gives rise to much less wastage of hay or silage than do normal racks. Combination racks are also an advantage where slatted floors are used as they reduce the amount of dropped hay or silage which would otherwise tend to clog up the slats and prevent their functioning properly.

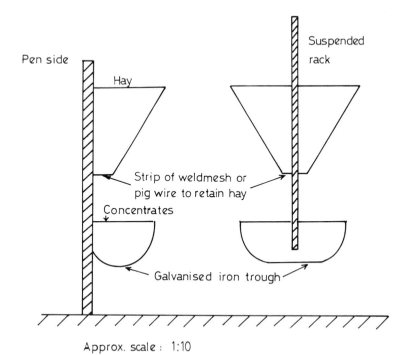

Fig. 10.2 Two types of feeding rack.

The trough space needed per ewe is about 0.30–0.40 m depending on the size of the sheep, while for independent hay racks 0.15 m is enough. Where sheep have continuous access to silage about 0.08 m of space per animal is required.

WATER

To animals on dry feed such as hay and concentrates only, a continuous supply of *clean* water is essential. The accent must be on 'clean', and care must be taken to ensure that the water tanks are kept clear of hay and similar material. Sheep are much more particular about the water they drink than are cattle and swine.

The usual cause in the interruption to the water supply is freezing and this must be guarded against, especially where ewes with lambs are concerned. Cutting off the water supply to these animals for any length of time means cutting off the lambs' milk ration. In countries of moderate climate a careful lagging of pipes will probably suffice, but in areas of low winter temperature such as much of North America electric heating will need to be supplied to drinking troughs. Finally, tanks and troughs should be placed in such a position that any overflow likely to occur is directed by gravity to the outside of the building and does not soak through a large quantity of litter. The sheep must start with a dry bed and it must be kept that way.

HOUSE SIZE

Having listed the prerequisites for a successful sheep house, the point must be underlined that the size of the building must be sufficient to accommodate a substantial number of sheep otherwise it will be uneconomical in the use of labour. The cost of carting fodder to a series of small buildings and providing the buildings with services can soon outrun the cost of erecting a new building.

The selection of a sheep house

BUILDINGS WHICH CAN BE CONVERTED

The flockmaster, having decided to house his sheep, has next to come to a decision on what sort of house to employ. In some cases the logic will be in reverse: the farmer, finding himself with a suitable building, may decide to house his sheep.

The best approach to the problem is to look over the steading and see if there is an unused or underused building which could, with modification, be made to accommodate sheep.

One of the most useful buildings is a Dutch barn; indeed, on many farms the Dutch barn has been used for decades for lambing ewes. These barns require little modification provided the long side with the most southerly aspect is open (this, of course in the northern hemisphere).

Cattle courts or fold yards can also be used. In many areas of England the farm buildings are roofed with pantiles. Although these may look too low-roofed for sheep, the air passes freely between the tiles and ventilation is not the problem it might appear. Strangely enough, old-fashioned broiler chicken houses function very well provided the windows are removed. The word 'removed' is used advisedly as if the windows are only opened some person being 'kind' to the sheep will shut them at some ill-judged moment.

The buildings which are really suspect for sheep housing are high, single-storeyed ones such as old-fashioned corn barns; being solid walled up to the eaves and usually with few or no windows, devising satisfactory ventilation is most difficult. Their height also creates the problem that as the hot air rises it is not removed but cools and falls back on to the sheep giving movement of air but little change.

If there is any choice in respect of the situation of a building for housing sheep or, indeed, any other livestock, a building on rising ground should be chosen. Where a new building is being erected it should be placed on the higher ground of the steading area. This positioning of the sheep house helps with ventilation and its effect is noticeable especially in foggy weather.

NEW BUILDINGS AND THEIR CONSTRUCTION

If no suitable building that could be modified to house sheep can be found on the farm, consideration must be given to the design of a new building. As previously noted, the amount of money that can be spent on a house for wintering sheep is limited. If, however, the cost can be shared with some other enterprise the situation is made much easier. Examples of complementary uses to which a building can be put are as a grain or potato store from harvest to mid-December. An alternative is to use the building for fattening turkeys up to Christmas and to house the ewes from then on. In a similar manner, the house could be used throughout the late summer and autumn for rearing beef calves. These types of activity require solid floors and are mainly centred on arable areas where there is ample straw for bedding. In all grass areas slatted floors may prove more economical than using bedding and here the building could be modified for barn hay drying or, in the wetter areas, used as a shearing shed and wool store.

(a)

(b)

Fig. 10.3 Sheep house in Nova Scotia, built for in-wintering in a harsh climate. (a) Outside. (b) Inside. (Farmer: Bruce Blacklock, Cape John, Nova Scotia.) (Photograph Fundy Films)

In the planning and construction of these dual purpose buildings, care must be taken to ensure that the basic requirements of both operations are met. A compromise which prejudices either enterprise does not make for overall success. Figure 10.3 shows a sheep house which was erected by farm labour and hence very economical.

Typical points to be considered are as follows. Where, for instance, the house is to be used as a grain store, the strength of the walls must have reference to the height to which the grain is to be piled. Where grain or potatoes are to be stored, the floor must be strong enough to withstand tractor traffic. Cross ties must not be so low as to impede the use of fore-loaders and similar implements. The floor space should be kept clear of stanchions and similar obstructions.

The position of the sheep pens should be planned out before the floor is laid and post holes for the major structures provided. When the pens are dismantled the socket holes can be blocked up with old sacks, a practice which is much more economical and effective than using fancy plugs of wood or metal.

The buildings which have just been described will have brick or concrete walls, corrugated asbestos roofs, sliding or roll-up doors of wood or metal as appropriate, and will be costly. The long axis of the building should, where practicable, run north and south.

The farmer who cannot devise a method whereby his sheep share housing costs with another enterprise is forced back on to something much less sophisticated.

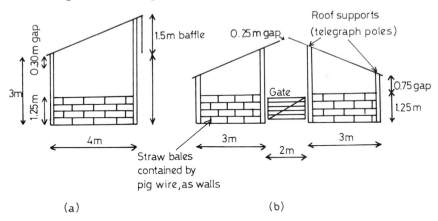

Fig. 10.4 Sheep houses: diagrams of end elevation of single and double span types. Pen depths are a minimum for convenience. (Approximate scale: 10 mm = 1 m.)

Some of the most successful sheep houses in Britain known to the author have been no more than pole barns built with second hand telegraph poles and having galvanised iron roofs, the houses having no walls other than those provided by straw bales. The straw bales are then retained by chain link fence. The usual design of the building is a rectangle with a double pitched roof, a central passage for feeding and the pens divided by wooden hurdles, the floors being made of rammed chalk, beaten earth or brick rubble (see Fig. 10.4). This type of house is essentially one for an arable area as the use of straw is high. Water supply can prove a problem in severe weather as the pipes and trough regularly freeze up.

Sheep thrive under the conditions which prevail in such houses and pneumonia rarely becomes the problem it can do in more sophisticated houses. There is one great danger however, and that is bacterial infection, especially such diseases as joint-ill. To guard against such attacks the straw walls should be replaced yearly and the floor thoroughly cleaned. When the house is bedded down the littering should be liberal.

Turning now to solid-floored, permanent-walled buildings, the use of straw need not be excessive provided it is kept dry. The build-up of dung is such that in a 90—100 day period no removal is necessary until the sheep go out to grass. The amount of straw required per ewe is about 50 kg.

Slatted floors

These have been in use in Scandinavia and Iceland for a long time and they would seem an obvious choice for anywhere where bedding is in short supply, especially if wood is cheap. The essential requirement in the construction of slatted floors is that complete freedom from underfloor draughts must be achieved. There must be ample room beneath the slats to ensure that no cleaning out is required during the housing period and a clearance of 0.8 m is adequate.

The bearers of the slatted grates must be properly supported on brick, concrete or properly treated wooden walls which are so constructed as to render the dung pit vermin-proof. The floor sections should be of such size and weight as to be readily moveable by two men. With regard to spacing and size of slats, a wide variety of measurements have been found satisfactory. In Norway the popular size is reported as 50—100 mm × 50 mm with a 25 mm gap. A popular size in Britain has been 37.5 mm × 37.5 mm tapered to 25 mm on underside width, and a 12.5 mm—20 mm gap. 60 mm × 25 mm hardwood slats with 20 mm gaps have also proved satisfactory.

The false economy of using inferior wood should be avoided. The timber should be of good quality, straight-grained, knot-free and properly milled. Knots and rough splintery edges on the slats will lead to accidents, damage to the feet and wool being plucked from the sheep by splinters.

An example of an actual building which has given good service over twenty years and which is still in first class condition has the following relevant features. The slatted floors are carried on wooden balks 150 mm X 100 mm. The gratings which are 2.25 m X 0.6 m were comprised of 50 mm X 25 mm slats at 25 mm spacing on three cross battens of 75 mm X 37.5 mm. The depth of the dung pit to the slats is 0.90 m. The timber balks are carried on brick piers as is the building which is of timber construction. The height of the pen sides is 1.0 m. The sheep are fed from combined hay racks and concentrate troughs mounted on the pen sides nearest the central gangways (as illustrated earlier in Fig. 10.1).

The timber is all softwood and the only problem arising from this construction is that occasionally a slat warps. This is dangerous as a sheep may catch its foot in the resultant gap and tear its hoof off in an endeavour to escape. A watch must be kept for such problems and warped or cracked slats replaced immediately.

Finally, sheep can be temporarily housed in such constructions as canvas marquees or frame buildings clad with polythene sheeting. These are finding favour with some farmers as suitable places in which to lamb a flock.

The management of housed ewes

FEEDING

The farmer, once he has adapted or built a sheep house, must make arrangements for the feeding of the flock throughout the housing period.

The most easily handled basic food which is most acceptable to sheep is high quality hay. The need for quality in hay cannot be over-emphasised; it must be made from relatively young herbage and must be well won, especially when it forms the major component of the ration. Musty or weathered hay should never be offered to in-lamb ewes as, in addition to its unsatisfactory nutritional status, fungus spores are liable to promote lung disorders.

The average intake of hay over the housed period will be about 1.5–1.7 kg per day or 150 to 175 kg for the hundred day period. In addition, some 10% would need adding for wastage, giving in round figures one tonne of hay per six ewes. Where grass silage is fed, a daily requirement of between 4 and 6 kg should suffice. The silage needs to have a dry matter content of at least 22% and have a high digestibility, the requirement for the period being about 0.50 tonnes per ewe. At the commencement of the indoor period hay or silage will suffice until about eight to six weeks before lambing when concentrate feeding should start. To this end, the ewes should be divided up in accordance with their colour marking at tupping. It is also desirable that the gimmers be kept separate throughout the winter period as this cuts down bullying. Dividing the flock up ensures that each group receives the amount of concentrates appropriate to condition and state of pregnancy. Overfeeding can be almost as harmful as underfeeding.

The ration should contain about 14% of crude protein provided by such foods as soya bean meal, whitefish meal or other protein-rich food, depending on price. The basic portion of the ration should be a cereal such as rolled barley or oats; dried sugar beet pulp can also be used. This concentrate ration should have 3% supplement of a mineral mix plus Vitamins A and D, and in many situations Vitamin E is also added. Feeding should start at about 120 g per head per day of concentrates working up to 0.75 kg depending on the condition of the ewe, the total quantity needed over the period being about 25 kg. About ten days before the first ewe is due to lamb, calcined magnesite should be added to the ration at the rate of about 35 g per kg to forestall hypomagnesaemia.

GENERAL MANAGEMENT

Before entering their winter quarters the ewes should have their feet examined for foot rot and should be put through a footbath containing a 5% solution of formaldehyde or a 4–8% solution of copper sulphate. It is an advantage if an arrangement can be made whereby the sheep can be walked through a water- and pebble-filled bath prior to treatment. If this cannot be done ensure that the sheep go into their foot treatment with their feet as dry and clean as possible. This is to ensure that the dip can get to the skin unimpeded. The animals should be moved slowly through the bath so that the chemical soaks the hoof thoroughly. The sheep should be dry when introduced to the house and every endeavour should be made to keep them and their bedding dry.

At the start of the housing period the sheep will require little attention, the main point for observation being ventilation. Any indications of problems arising, such as finding the sheep with their backs wet with 'dew' should receive attention as this is a sure sign of poor ventilation. Always remember it is very easy to give sheep too little ventilation. Ventilation being under control, all the sheep need is a foot check and a periodical walk through the footbath.

LAMBING

The most obvious difference between indoor and outdoor lambing is the question of mis-mothering. Outdoors, the ewe seeks a spot away from other ewes and there proceeds to lamb and mother her litter. There is nowhere she can find seclusion indoors with the result that indoor lambing requires constant supervision.

The shepherd should construct a series of lambing pens either in a section of the main building or, preferably, in a smaller adjacent house. Immediately a ewe starts to lamb she is removed to a lambing pen where she and her lambs remain for about 48 hours. This gives time for the ewe-lamb bond to be properly developed. The family is then ready for turning out into a yard or small field with other families.

The ewes of the breeds and crosses used for intensive meat lamb production have strongly developed maternal instincts – otherwise they would not look after sets of twins and triplets as is required of them. They are therefore prone to steal lambs. This enticing of lambs can develop into a tug of war between the mother of the lambs and one or more unlambed ewes. The lamb becomes confused and the mother may let it go, especially if she is a gimmer, the result being that the lamb gets no colostrum and is probably lost.

COLOSTRUM

Colostrum is very important in respect of large litters. The smallest lamb in a set of triplets is liable to receive less than its fair share unless the shepherd is very attentive. To ensure a satisfactory availability of colostrum, ewes with a single lamb and a large initial milk supply should be milked out by hand and the colostrum stored in deep freeze against the time a ewe is short of milk. The thawed-out colostrum can be administered to the lamb by means of a stomach tube. Narrow, blunt-ended tubes with an aperture at the side can be bought from veterinary suppliers, together with small bottles which hold about 90 grams. These bottles or glass cylinders are open at one end, the other end tapering to a spigot on to which the tube is fixed. The tube is passed down the gullet and the colostrum run in under gravity. Care must be taken to see that the tube arrives in the stomach and not the lungs. In some cases the colostrum may be thick and somewhat viscous making it difficult to administer, but this can be overcome by diluting it with warm water.

Using the above method the shepherd can ensure that each lamb gets sufficient colostrum at the right time. It must not be forgotten that the ability of the lamb's gut to absorb the protein of the colostrum declines from the first feed and not with time; that the first feed is of a sufficient quantity is therefore important. The ewes with their litters are held in the yards or paddock until it is judged appropriate to introduce them to the intensive grazing system. At this time of holding in yards, a supplementary diet of succulent food will find favour with the ewes and seems to have a beneficial effect on their milk yield. Such 'roots' as swedes, mangels, fodder beet or potatoes are all found acceptable.

Housing of sheep other than ewes

HOUSING OF EWE HOGGETS

In its simplest form this makes use of stone-built, two-storeyed buildings such as are found in the Yorkshire Dales. These houses are usually set away from the steading in a field with a water supply in the form of a spring or stream. The hay is stored in the loft while the hoggets are housed below. They may be let out only to drink or they may wander in and out at will.

As will be discussed in Chapter 11, hill farmers with little or no enclosed land have this problem of hogget wintering; some farmers are moved to consider housing the hoggets in slatted floored houses and buying in hay to feed them.

The type of house required is exactly the same as for ewes, but the floor space is, of course, less: 0.40 m² for small breeds such as Welsh Mountain and up to 0.50 m² for Scottish Blackfaces. A clearance between the slats and the ground of 0.40 m will hold a winter's dung comfortably.

The major points which need attention are as with ewes, good ventilation and, what is apt to be forgotten with non-lactating sheep, a steady water supply. Winter-fattened hoggets can also be housed. Indeed, some British farmers keep them in courts in the same way as beef cattle are kept. Ventilation in these courts or covered yards is seldom a problem. The major difficulty to be resolved is to devise a feeding system which will enable one man to look after a large number of sheep.

Intensive grazing and housing having been dealt with we will now return, in the next chapter, to the original source of most of our sheep — the hills and marginal land.

Points to remember

1. Reasons for housing sheep.

(a) Predators in some areas.
(b) In areas where sheep are milked, housing overnight makes the job easier.
(c) Harsh snowy winters as in Scandinavia and N. America.
(d) Reintroduction of grass breaks on arable farms. Only intensive methods are economic for sheep farming on expensive land – hence housing.

2. Factors on which economical winter housing depends.

- Pasture protection.
- Nutrition.
- More efficient utilisation of labour.
- Health (particularly prevention of perinatal deaths).

3. Disadvantages.

High capital cost (therefore seek an alternative use for the building).

4. Basic requirements of a sheep house.

- Good ventilation.
- Adequate floor and trough space.
- Suitable source of daylight.
- Adequate clean water.
- Dry lying.
- House must be large enough to accommodate an economic number of ewes.
- Area per ewe: 1.2–1.5 m² on solid floors and 0.8–1.0 m² on slatted floors.
- Trough space: 0.30–0.40 m per ewe.

5. Slatted floors.

Suitable when bedding is expensive but:

- Exclude under floor draughts.
- Allow enough room for accumulation of dung.
- Make vermin-proof.
- Avoid rough-sawn timber.

11 The exploitation of hill and marginal land by sheep

Before the late eighteenth century the mountain areas of much of Europe were exploited by what is known in Scotland as the *shieling system*.

Under these systems the rural communities moved their livestock, which was mainly cattle with a few sheep and goats, to temporary homes on the high ground in the springtime. The men returned to the lowlands for a short period to plant the spring cereals. Throughout the summer the cattle would be herded by the children on the hill grazings, some of the cows being milked and cheese and butter made.

The township families would return to their home base at the end of summer for the low ground harvest. Practices varied from country to country. In some areas hay was made on the high ground depending on weather conditions, the opportunities for hay-making being infinitely better in the Alps than in the West Highlands of Scotland.

The chief agricultural significance of this system is that the hill carried a relatively heavy stocking in the summer and, being under the control of herders, the pasture would be eaten down in a fairly uniform manner. This is quite different from what happens in many such areas today, where over-grazing in winter and spring are commonplace while in the summer the herbage gets beyond the control of sheep.

The original wether flocks which first occupied the higher hills of Scotland after the crofters were evicted have now been displaced by breeding stock. These mature wether flocks were run for some years on the hills, being shorn annually and slaughtered at 3–4 years of age. The wether flocks disappeared because the low price of wool combined with the growing demand for lamb instead of mutton made them uneconomical. Their place has been taken by ewe flocks and now all the hills of Scotland, as do the hills in the rest of Britain, carry ewe stocks.

The new system has disadvantages as well as advantages. One

important asset is that in set stocking with indigenous sheep on unfenced hills, the sheep are hefted or 'bound to the ground'. This ensures a minimum of straying in spite of the absence of other restraints. The second important advantage is that in tick-infested areas the sheep develop a substantial resistance to tick-borne diseases. The two serious disadvantages of the system are that the hills tend to be (a) understocked in the summer and (b) overstocked during the winter. This state of affairs leads to a general degeneration of the sward.

The general accumulation of uneaten dead herbage is slow to decompose, as is usually the case under acid conditions. Acid soils also tend to retain the essential element phosphorus in a form which is less available to plants than in soils with a normal pH. Herbage that has been broken down by passing through the digestive processes of grazing animals provides appreciably more available nitrogen and phosphorus for growing plants than can arise from the normal decomposition of plant litter in an acid environment.

In addition to the limited availability of some minerals there is also the point that each crop of lambs and wool removed from the hill with no fertiliser application being made must mean that the soil becomes impoverished.

The length of time spent by the ewes on the hill varies. In some areas of Scotland the ewes never leave the hill except at gatherings whereas in many areas in Wales the ewes come down from the hills for tupping in late October-early November and do not return until lambing is finished in May. In other cases the ewes may stay on the hill until January when they are brought down onto enclosed land. The routine management of the hill farm will naturally depend on the type of holding, local practices and so on. To give particular examples, some farms will be set stocked with a bound flock on a hill with no fenced boundaries except in the main valleys, and the land divided into *hirsels* (which are the areas normally handled by one shepherd). Each hirsel will be divided into subsections or *hefts*, usually bounded by the streams and dividing ridges of the area. Many of the farms will have only enough enclosed land to provide grazing for the tups, a few ponies and a house cow. Others will have enough for some hay to be made and, possibly, enough land to feed some of the wether lambs on rape. In some areas such as the north of England the high fells are held in common by the farmers in the dales. Some of these commons are stinted while others are not. The normal practice on such farms is to withdraw the sheep from the hill to the enclosed fields for the winter. In Wales many farms have relatively large areas of enclosed mountain known as *ffridd* where the sheep are kept over the winter months.

Having made a rough sketch of some types of hill land organisation in Britain, we now need to look at the farming in a little more detail.

The major characteristics of hill farms

The brief look at hill farms we have taken shows that even in a small country there can be a great diversity of practices. There are also similarities which we must now examine.

In Britain the hill grazings are generally characterised by the following:

- Sour peaty soils, lacking in fertility and low in essential minerals, especially the bone formers calcium and phosphorus. What minerals are present are often of low availability.
- In summer the weather is often wet and cool. The winters tend to be wet and cold with heavy snow falls at times and what winter sunshine there is is often obscured by cloud – therefore there is no growth of herbage. On north-facing situations the start of spring growth is even later and the end of summer production earlier.
- The bulk of the plant population consists of late-starting, slow-growing perennials many of which are tough, woody and not readily eaten. In summer and early autumn there is usually plenty of keep but in winter, and especially early spring, there is very little food.

Dominant vegetation

There are many varied types of hill grazing but in Britain the main types are as follows, being dominated by heather, mat grass or bents.

HEATHER (*Calluna* spp.)

A hill which is covered mainly by heather has a very low stocking capacity. Sheep can utilise up to 40% of the current season's growth of heather without affecting plant productivity. Above this limit growth is restricted and heavy autumn grazing is particularly damaging. The nutritive portion of the heather is the young growing shoots and for this reason heather should not be allowed to get too big and rough but should be burned regularly in rotation. Heather is, however, a slow-growing plant and overburning will result in a weakened plant which will ultimately die out and be replaced, often by a worthless species such as nardus. Generally speaking the rotation should be 15–20 years. If a heather hill is very rough it is a mistake to burn off in one operation. The burning should be done in sections so that a proper succession can be established in the regeneration. Great care should be taken to keep heather fires under control, and to this end heather or other herbage should not be burned in a high wind. If burned in a strong wind the fire will go out of control, burn deep and destroy the substrate, and the next problem will be erosion. This is particularly the case on steep slopes where scree has been stabilised by herbage. Evidence of injudicious burning can be seen on most hill lands.

Burning is best done under conditions of light wind and when the heather and soil are not too dry. The burning should also be carried out on limited areas over a definite time scale. If these restrictions are observed a sudden change in wind speed or direction need not prove as dangerous as often would be the case when half a hill side has been set ablaze. In Britain and other countries there are legal restrictions on heather burning and it may only be carried out at prescribed times.

The object of the burning is to clear the thickest of the rubbish and let in the daylight so that the old stools are induced to throw off new shoots and seeds allowed to germinate. Heather, however, is not particularly nutritious and, at the best of times, will only provide maintenance.

MAT-GRASS (*Nardus stricta*) AND FLYING BENT (*Molinia coerulea*)

Nardus is a hard, wiry grass which is a poor producer and is only palatable for a short period during early summer. The treading of stock tends to eradicate nardus and grazing with cattle helps in getting rid of it.

Molinia is called *flying bent* because it is deciduous, the leaves blowing off the plant in winter. It provides some keep in the summer but little in the winter unless made into hay.

BENTS (*Agrostis* spp.) AND FESCUES (*Festuca* spp.)

These are by far the most productive grasses of the high ground; in association with wild white clover they can produce a pasture which will fatten stock. The fine-leaved *fescues* are typical of calcareous soils such as the English Downs, while *agrostis* is the dominant grass on the good grass hills of northern Britain. These *agrostis* pastures give a marked response to lime and phosphate where these can be applied, and it is towards this type of pasture that improvements should be directed.

OTHER CONTRIBUTIVE PLANTS

Various plants other than grasses make a contribution to the food of hill sheep, while there are other herbs and shrubs which are undesirable. Of the useful plants, bog cotton (*Eriophorum vaginatum*) is one of the best known. It is a sedge which, in Scotland, is known as draw moss. It makes an important contribution to some flocks in the early spring; unfortunately it is also a plant of soft, boggy ground such as harbours the host snail of the liver fluke.

Low-growing shrubby plants such as bilberry (*Vaccinium myrtillus*) are also grazed by sheep, especially in their young stage. Wild white clover (*Trifolium ripens*) makes an important contribution where soil and grazing conditions are such as to encourage it. Of low shrubs which are undesirable, whins or gorse (*Ulex* spp.) and, on the lower slopes, bramble (*Rubus* spp.), are amongst the most important in Britain. They are both mass colonisers which make little contribution in the way of food and tend to sterilise large areas as far as sheep are concerned.

Other undesirable plants are deer grass (*Scirpus* spp.) which is really a sedge and bracken (*Pteridum* spp.) and should, where possible, be eradicated. Bracken is particularly objectionable because it colonises the deep soils of the lower slopes. Not only does it shade out good grasses and clovers but it hides sheep and makes them difficult to locate and to gather. This was particularly serious in the days when there was no good anti-fly dip: struck sheep retired to the bracken patches and died.

Hoof cultivation by cattle helps to eradicate bracken, but under some conditions in some seasons of the year it can be poisonous to cattle. Bracken can be controlled by repeated cutting and crushing. In some areas where boulders do not present a problem a set of disc harrows pulled by a crawler tractor makes an excellent job. Aerial application of chemicals is also effective but tends to be expensive.

Basic hill management

The factor which has the major influence on the management of hill sheep stocks is the marked seasonality of herbage growth. In some areas the season is particularly short, but the hill lands of Britain do not usually show the midsummer drop in herbage production characteristic of the low ground. The sheep is the ideal animal for coping with this form of production as the demand curve of ewes and lambs for food is closer in fit to hill land growth of herbage than that of other farm animals. The summer growth is relatively abundant and the lambs are there to consume it. At the end of the summer all but the breeding stock are removed from the hill and the wether lambs are sold, as are the ewe lambs surplus to breeding requirements. Indeed, in most cases the stock ewe lambs are removed for away wintering. The removal of these animals does not, however, ensure that the breeding ewes will get through the winter adequately fed. The problem of balancing the winter requirements of the ewes with the summer production of herbage is central to the management of hill land. A similar problem arises in semi-arid areas with regard to rainfall. In these areas unnaturally long droughts cause the stock to run short of food while, conversely, in hill areas of countries like Britain snow fall can be so heavy and persistent, with temperatures so low and so far into the spring that ewe mortality from malnutrition can be substantial in either case. Ewes can also be so weakened by these conditions that they die at lambing or in early summer.

THE PROBLEM OF WINTER FEEDING

There are many devices which flockmasters use in their endeavours to resolve the problem of winter feeding. The first factor to be resolved is that of stocking rate and the type of stock employed. The sheep must be appropriate to the ground. One cannot be successful if a large breed of ewe at a heavy stocking rate is used on a poor, thin soil which is low in nutrients.

While mentioning stocking rate, a reminder should be given that while the stocking rate on most hills is low, the stocking density can, at certain times, be very high with resultant problems. These difficulties can arise in such places as along spring lines where the underground water brings up minerals and is also at a temperature that promotes the first growth early in the year. This leads to overgrazing (unless steps are taken to prevent it) and the encouragement of worm parasites. Another source of trouble is supplementary feeding when always given in the same place. Finally, fertilisers and reseeding treatments provide an excellent situation for the proliferation of stomach worms unless some element of control is introduced, as the sheep will tend to congregate in these areas.

Let us look at the simplest situation first, where the sheep are set-stocked, remaining all the year on the hill and lambing there. Little supplementary food is provided except in very adverse weather conditions. How supplementary feeding will be carried out depends on the overall system of utilising hill and mountain pastures. These vary widely. The simplest, common in many mountain areas in Scotland, is that the stock remains on the hill throughout the year. In other systems, such as are common in the Pennines, the sheep are brought down to grass fields for the winter and early spring. In Wales the sheep are withdrawn to the *ffridd*, which is enclosed land at a lower elevation but not usually up to the same quality of rotational grass land. The date when sheep are brought down to the low ground may be as early as late November or as late as the end of January. In some areas such as north-east Scotland the sheep may spend some part of the day in winter on arable crops, and this also applies in some parts of Devon. Again, in other areas the sheep may come in to self-feed silage clamps to augment their grazing.

Where the sheep remain on the hill an effort is made to shepherd the ewes to the high ground from weaning time onwards, thus preserving a surplus of summer growth on the lower slopes for winter consumption. In other words, it is a way of providing *foggage* (i.e. grass which has been left almost as hay but harvested *in situ* by the grazing animals).

Supplies of hay are usually cached at strategic places as well, for use in times of adverse weather. In some cases feed blocks are provided. These blocks are made of maize or other cereal meals bound together with molasses. Urea is also incorporated as a protein substitute. The usual rate of allocation is one 25 kg block to 40 ewes per week.

Concentrates can also be fed in the form of cubes or whole grain. Feeding meal or small grain requires the use of troughs and on a large sheep walk this is not very practicable. The maize or other concentrates are fed in quite small quantities, say 90–120 g per head per day. Such feeding is not normally started until eight to six weeks before lambing except under severe weather conditions, the object of the exercise being to prevent the ewes from losing weight. Hay can also be used and at one time was the only supplement.

Quantities used vary but about 4 kg per week is a realistic figure. The big problem with hay is the difficulty of carting it onto rough and steep areas.

Three points need to be emphasised concerning supplementary feeding:

- Sheep need to be educated to eat such things as grain and concentrates. An adult ewe presented with a handful of maize for the first time may completely ignore it even though she is desperately hungry.
- When ruminant animals are really short of fodder — and this applies not only on the hills and in range conditions — it is better to give two or three reasonable feeds of hay per week rather than a small one every day. The logic of this is that with small feeds the 'boss' animals get the lion's share of the food by bullying their inferiors. With the larger feeds the bullies sate their appetites before all the fodder has run out leaving the shyer and less forceful animals a share.
- Where one is feeding fodder or concentrates outdoors on the ground, it should be spread in a circle or circles. This ensures that the animals move more rapidly round and round until the food is cleared up and by this means the shy animals are not so readily driven off. This makes for a more equitable distribution of the food than happens when it is thrown out in heaps.

The previous examples from different parts of the country will suffice to show that the ways in which hill sheep can be fed and managed are diverse and what may be anathema to sheep men in one area may be accepted practice in another. What the individual farmer needs to do is to make use of his basic knowledge of sheep, their physiology and nutritional requirements and accordingly devise a system that not only suits his own particular circumstances but also makes economic sense.

In conclusion, it must be mentioned that although it may seem self-evident to provide supplementary feeding to hill sheep in one form or another, especially when conditions are severe, there are some shepherds who show little enthusiasm for it. Such people claim that the sheep are made 'soft' by feeding, that they will not forage and that they spend all their time waiting for the food trailer. However, feeding would need to be much more than supplementary for this situation to develop, as normal supplementary feeding does not provide a full gut for long! In spite of the additional nutriment that supplementary feeding provides, the ewes will be keen enough to graze. Even those satiated with big meals twice a week will work hard enough on the other five days.

Having made these points on winter feeding, it must not be imagined that a lavish outlay of supplementary food will ensure success in hill farming. As with any other form of farming, the result will be failure if the cost of inputs exceeds the value of the outputs! The hill farmer needs to consider carefully the type of food he provides, the quantities, the method of feeding and the timing of operations for his particular holding. A régime which is satisfactory on one farm in one area may not be appropriate for a similarly sized farm elsewhere. Also, the farmer must bear in mind that a programme which is successful in one year may prove disappointing in another. In a vocation like hill farming where the elements often ensure that one season is quite different from the next, the flockmaster should develop a flexible approach and, like a good general, keep his options open as far as possible.

Hogget wintering

One feature of the management of a hill sheep farm that has a major influence on the stocking rate and on supplementary feeding is the treatment of the ewe hoggets.

In the most primitive systems, the lambs were left with their mothers until they weaned themselves and they went through the hogget and gimmer stages of their lives without ever having been off the hill. The advantage of this was that they did not suffer any weaning check which lambs normally experience. It also meant that the family tie on which hefting is based was strengthened.

The big disadvantage of this kind of system was that the number of ewes that could be carried was very much reduced. There was also the problem that in a severe winter the hoggets suffered a setback from which they never really recovered even to the extent that their small size was reflected ultimately in their cash sale price. Furthermore, poorly reared ewes tend to have a curtailed lifetime's production of lambs.

This problem of poorly grown females can be avoided by keeping them for another year before putting them to the tup. This delayed mating is achieved by sewing a piece of sacking over the female's rump, the sewing twine being worked through the fleece and tied.

Most hill sheep walks are incapable of growing lambs to a suitable size in their first winter. What enclosed land hill farms do possess has usually carried so many sheep during the rest of the year – tups, ewes with twins etc. – that the enclosed land tends to be *sheep sick* and no place for growing stock. (*Sheep sick* is a term applied to areas of land which have carried too many sheep over an extended period of time and have become heavily contaminated with stomach worm larvae, foot rot bacteria, and other pathogenic organisms.)

The problem of what to do with hoggets over the winter period has been solved in a number of ways. One early method, now obsolete, was to have a separate hill or section of a hill away from the breeding flock which was more fertile, nearer at hand and more readily herded than the main hill grazing. Here the hoggets were run until the make up of the breeding flock in the following year. A place name common throughout the Scottish Borders, Hogg Knowe, is almost all that remains of the system.

Most hill farmers with too little enclosed land solve the problem by *away wintering*. Here the weaned lambs are sent away to a low ground area, usually a dairy farm, the sheep farmer paying so much per head per week to the dairyman. This practice is known as *agisting stock*.

The demands made on the low ground farmer who takes in the hoggets are not onerous. He needs well fenced fields and someone to look after the sheep – 'give an eye to' – is the appropriate phrase as such animals require little attention. The stocking rate should be light both in the interests of the sheep and the farm. Two or three animals per hectare is a satisfactory rate.

All that hoggets require is pasture, plus a little hay during hard weather. Overfeeding is undesirable – the animals need to be well-developed but not fat. In passing, it should be said that baring down cow pastures by sheep during the winter has a stimulating effect on the growth of grass in the following spring, provided the grazing has not been severe or too protracted. This is where the major difficulty arises in the case of hoggets away wintered on a dairy farm. The farmer wants to get rid of the hoggets in late March or early April; this means that the hoggets have to go back to the hill when there is yet not enough fresh grazing for the ewes, let alone the hoggets. The most satisfactory solution to the problem is for the hill and low ground farms to be in the same occupation. This, however, is not normally the case.

The costs of away agistment get higher and higher each year, both in terms of rent and of transport, as the farmer's costs over the past few years have risen much faster than his income. Increased fuel costs have made a serious impact. As these costs have risen, hill farmers have of necessity reassessed the pros and cons of away wintering and some have turned to housing their hoggets. This was mentioned previously in Chapter 10 when discussing sheep housing, and the houses described in that chapter are suitable for this purpose.

The requirements of hoggets are less than for ewes in terms of area and trough space. With regard to feeding, it must be emphasised that the objective is that the animals should grow steadily but not get fat. To achieve this state a ration of about 1 kg per day of hay with 100 g of concentrates should suffice. These figures are a general guide and what the shepherd needs to do is to regulate the quantities fed in relation to the size and condition of the animals and the growth being achieved. It is an advantage if the hoggets can be provided with an outdoor area for exercise. Once again the reader is reminded that to make a success of the housing adequate ventilation is absolutely essential.

There are a good few hill sheep farms in Britain that could be made much more productive and where provision could be made for out-wintering hoggets from the farms' own resources. These farms may have areas of land which are not too steep and have a sufficient covering of soil to respond to fertilisers and cultivation. Such areas can have their production raised substantially and could be a source of hay for winter feeding and a pasturage for ewe hoggets.

This brings one to the question of hill farm improvement and, while there is insufficient space to go deeply into the subject we must mention some of its major aspects.

Hill farm improvement

The major weaknesses of hill farm production generally tend to be:

(a) a low lambing percentage
(b) high neo-natal lamb losses and
(c) poor lamb growth leading to lower than desirable weaning weights.

The principal factor which causes this state of affairs is the poor nutrition of the ewes in winter and early spring.

Britain is fortunate in respect of knowledge on hill farm improvement thanks largely to the efforts of the Hill Farming Research Organisation (H.F.R.O.) which was set up some 25 years ago. This organisation started its research from the standpoint of the unfenced Scottish sheep farm where the sheep resided on the hill for all or for the major portion of the year. It was speedily realised that this traditional management created a vicious circle with the stocking rates in the summer being set to enable the sheep to gain a certain minimum winter food supply. This gives rise to selected grazing with its ensuing problems.

It was also appreciated that sheep production had to be considered on a year round basis and that any action taken at one point in the cycle would promote a reaction in another. The accumulated evidence from studies in various aspects of hill management improvement pointed in one direction — the need for some control of grazing. The type of hill grazing which showed by far the greatest potential for improvement was the agrostis-fescue sward. Field experiments over a long period of time led to what has come to be called the *two pasture system*.

THE TWO PASTURE SYSTEM

Under the two pasture system, an area of improved grassland is incorporated into the hill farm. In the case of agrostis-fescue pastures, this improvement can be achieved by fencing and fertilising selected areas which are subjected to controlled grazing. In other areas such as nardus- and molinia-dominated swards, a process of reseeding has to be adopted.

By using the improved pasture during lactation, that is, from April to August, followed by a rest period and again grazing just before and after mating time, milk yields are improved, better lambs are obtained and the enhanced body condition of the ewes at mating ensures a good crop of lambs. The increased production obtained from a reseed which can be stocked at 14–16 ewes per hectare will allow an overall increase in ewe numbers. It follows that additional supplementary feeding may be required in late pregnancy while the ewes are on rough grazing. Increase in output of 160–200% has been achieved by the two pasture system. Whilst this sort of improvement is readily achieved on an agrostis sward without reseeding, under most other conditions reseeding is most likely to bring early success.

In reseeding, a number of points should be borne in mind. Acidity has been mentioned earlier and this has to be corrected before any substantial improvement can be made, but it must be remembered that it is possible to over-lime and lower the availability of cobalt, copper and other mineral elements. A pH of 5.8 is suggested by

H.F.R.O. as a maximum at which to aim. Improved pasture production in any hill area must involve the use of clover. While there are conflicting reports from various experiments on the effectiveness of inoculation, H.F.R.O. holds the view that on peat soils clovers should be inoculated with an appropriate strain of *Rhizobium*. Advice on this matter should be sought from such agencies as A.D.A.S. or college advisory staffs.

It will be appreciated that any form of improvement will require an outlay of capital, for fencing, fertiliser, seeds, additional stock and so on. This is often the most difficult part as after the improvements are established they will generate more capital. The point to remember is that *one should improve the best first*: in other words, start where the response will come quickly and in substantial volume.

The improvement and fencing of the new pasture necessary for grazing control also offer the advantage of a more efficient use of labour.

Having looked at improvement generally, a few words on the separate aspects of the work are appropriate.

HILL DRAINAGE

This can be undertaken on its own and need not be in conjunction with other improvements. Such drainage can make available land which would otherwise be dangerous to sheep; it helps to eradicate the mud snail which harbours liver fluke and also helps to eradicate worthless sedges. Hill drains are normally open drains ploughed out with specialised drainage ploughs, cutting along the contours to natural water courses. On no account should drains be cut down steep gradients or serious gully erosion will result. Hill drains do constitute a problem where cattle are grazed as the cattle tend to tread in the drains. This is especially the case on peaty ground and the result is that unless the drains are fenced frequent repair to the drains is called for, a job which is not to the liking of most shepherds.

FENCING

With the exception of bracken control all remaining methods of hill land improvement call for some degree of restriction on the movement of the grazing animals. Improvement of the land by fertiliser treatment and/or mixed grazing is not possible without fencing. If such improvement is attempted it will fail. Areas which have been fertilised must be fenced off otherwise they will be overgrazed at the expense of the rest of the holding which will become rougher, while the herbage of the treated portion will deteriorate and also suffer a serious build-up of parasites. In some areas of North Wales and Northern England there is also the problem of common grazings which makes improvements difficult.

On any area where cattle are grazed the main drains and any soft areas where animals may become bogged down should be fenced. As cattle do not appear to have a well-developed hefting instinct, a stock-proof boundary fence is also a necessity.

Space does not permit the examination of all the various sorts of fencing appropriate to different stock requirements but a number of general points need to be made. Effective fencing of any sort tends to be expensive so when permanent fences are erected the farmer should ensure that the job is done well. Most new permanent fences on hill land are of post and wire as the cost of the stone dykes of earlier days has become prohibitive. Here the main aspects to check are that the number of strainer posts is appropriate to soil type and topography, that the strainers are buried to a sufficient depth and that the sprag posts should be plated to prevent movement. The strainers and intermediate posts used should be pressure-treated with a good wood preservative. The staples holding the wires to the posts should not be driven fully home as one needs to allow for tightening the wire. The wire will need tightening within a year of the fence being put up and should subsequently be examined and retightened on a regular basis.

The normal height of a fence is about 1.10 metres, consisting of six or seven rows of plain wire topped by two rows of barbed. The lower wires should be nearer together than the higher, with the lowest wire about 100 mm from the ground with the following in the order of 120, 130, 130, 150, 150, 160, 160 mm between.

As for less permanent fences, the various types of electric fences should be considered. The fact that all forms of fencing can become drifted over in snow storms must be borne in mind and the farmer should not be taken unawares by this situation.

Finally with regard to fencing and other permanent capital improvements such as hill roads, the integration of farming and forestry can offer the opportunity to distribute costs between two enterprises and also provide farms with better access and livestock with sheltered areas. These areas provide not only storm refuges but also areas of micro-climate where growth may start a little earlier in the spring.

MIXED GRAZING

The continuous grazing of a pasture by sheep, especially ewes and lambs, usually leads to a deterioration of the herbage. Chalk downs are exceptional in that they can support grazing sheep alone with little difficulty. The removal of the wether flocks from the hills has accentuated the deterioration as these animals tended to eat down the rough herbage during the winter more effectively than ewes as they have lower nutritional demands than pregnant ewes and could therefore be stocked more heavily.

The continual selective grazing by sheep promotes the growth of shrubs such as gorse (*Ulex* spp.); while fire can be used to mitigate the problem the best results are obtained by grazing with cattle, which are more controllable. Various types of cattle may be used for the purpose. They can be store cattle of beef breeds or their crosses, or even dairy heifers. Some hills maintain herds of hardy suckler cows such as Galloways and West Highland. The author has found Ayrshire dairy heifers excellent animals for hill improvement.

How many cattle can run on the hills and for how long depends on the farmer's objectives. On some hill farms the beef enterprise may be as important as the sheep, but here we are merely concerned with cattle as land improvers for the sheep. Many hill shepherds are hostile to cattle, not only because they tread in the drains and occasionally knock down the dykes, but also more importantly, in their view, they deprive the ewes of their winter food. This is often fair comment, as there is a tendency to bring cattle on to the hills too late and leave them there too long. To be effective the cattle should be put on the hills relatively early in the spring. The cattle cannot compete with the sheep for the short bite and are forced to eat the rough herbage. As the new growth comes up through the old foggage, the cattle pull off both old and new herbage together and thus do a job of pruning which allows an unhindered growth of the new material.

If cattle are brought in late in the spring when there is a good growth of grass the cattle beast, being no stupider than other animals, will reject the roughage and join the sheep in eating down the fresh grass. The cattle should be removed from the hill in mid-summer in order that the herbage growth of late summer and early autumn may be preserved *in situ* for the ewes during the winter. The number of cattle a hill area can carry has to be determined by trial and error and will depend on the type of hill and its general topography. One cattle beast to ten or twenty ewes is a rough average, depending on conditions.

FERTILISER TREATMENT

Most hill lands are markedly deficient in calcium and phosphorus and no great productivity can be expected until such deficiencies are made good. The cost of applying fertilisers on rough land far from hard roads is usually very expensive. Because of the lack of accessibility in hill land, countries such as New Zealand have made progress with the aerial application of fertiliser and of pelleted seed, but in Britain little has been done so far. Because of the high cost of applying fertiliser, the areas for treatment should be selected with great care and only the best land where success is virtually assured should be treated in the early stages of improvement. This is a cardinal tenet of all types of farm improvements — resources should be applied at the point where they give promise of the best return — but this is especially true where resources are limited as they so often are in hill farming. Having decided on the area to fertilise, the next decision is which fertiliser. In most circumstances the materials which give the best response are ground limestone and basic slag. These counteract the acidity and the slag also provides phosphorus. For some reason which is not clear, wild white clover is particularly responsive to slag. Liming without phosphate gives little reward.

Where land has been fenced, fertilised with the above chemicals and appropriately grazed, the situation can arise on the lower slopes of the enclosed land that nitrogen will become a limiting factor in respect of herbage production. Where nitrogen is used on high ground care has to be taken that the applications are not made too early in the season, otherwise a lot of trouble can arise from late frosts devastating the soft, early growth promoted by the nitrogen. Any of the commercial nitrogen fertilisers may be used but nitro-chalk is probably the best because it does not have the acidity of sulphate of ammonia.

RESEEDING

The ultimate in most hill land reclamation is to try and make a substantial change in the constituent plants in the sward, or at least in their proportional relationship to one another. This can be done as we have suggested by fencing, fertilising and judicious grazing. To make radical alterations in respect of species will require reseeding to be done. This can be attempted by seeding from the air with pelleted seed, broadcasting by hand or machine, cutting the seed into the ground with a sod seeder or ploughing out and seeding either directly or following a pioneer crop or crops. The ultimate aim of improvement should be to establish good hill sward that can be maintained over a long period of time rather than an ephemeral lowland type pasture. It is quite possible to produce a sward of bred strains of perennial ryegrass and other improved grasses by dint of cultivation and heavy fertiliser treatment which look and behave well over the first year or so, only to revert to rushes especially on deep wet land.

The majority of seed mixtures used in hill land reclamation contain a large proportion of perennial ryegrass (*Lolium perenne*). Cocksfoot (*Dactylis glomerata*) for the drier land, timothy (*Phleum pratense*) and meadow fescue (*Festuca pratensis*) for the damper soils, and in all cases white clover (*Trifolium repens*), are also present as ingredients. Seeding rates are usually fairly high – about 40 kg per hectare.

It will be gathered from the above that reseeding is a very expensive business which should not be entered into lightly. Professional advice should always be sought from local experts on seed mixtures, fertiliser treatment and dates of sowing.

While there are areas of hill country which have been redeemed by the plough, there are others where success has been very limited and any such form of improvement should be approached with great caution as the ploughing up of poor, thin soil often meets with disaster. Where the hill soil is of good quality and normal and low ground farming techniques can be introduced and the land fenced, the area ceases in effect to be hill sheep pasture and becomes normal rotational farm land. Areas such as these can contribute to the improvement of adjacent land by an on/off policy of grazing; the stock grazing the enclosed land for a short period and then returning to the hills as described in the two pasture system. In this way the animals transfer some of the fertility via their dung to the hill. These enclosed areas can also be used for the production of hay and silage to augment the rations of the hill stock.

To sum up, what one is trying to do is to increase the amount of keep available, to control the stock in such a manner that enables it to consume the herbage at the appropriate stage of growth or, if it is not eaten immediately, to ensure that it is properly conserved.

As the late Professor Martin Jones repeatedly emphasised, what one tries to do from the botanical standpoint is to prolong the vegetative phase in the herbage as long as possible, while keeping the reproductive phase in check. This means heavy stocking in the late spring and early summer, and when the reproductive season of the plants begins to pass a reduction of the stock in order that foggage can build up for the winter.

This building up of foggage is the normal conservation method for high hill and mountain grazings, but where a marked improvement has been made in herbage productivity some mechanical harvesting will need to be done. Conservation of harvested material is best done as silage on a hill farm as harvesting equipment can be kept to a minimum – a tractor, a mower and a backrake. There is the additional point that the weather in hill areas is more amenable to the making of silage than of hay.

Health problems of hill sheep

Hill sheep do not suffer to any greater extent than low ground sheep from transmissible diseases, but one feature which distinguishes some hill farms from many others is the seriousness of tick infestation. On these farms tick-borne diseases such as tick pyaemia can be troublesome. The sheep stocks on these farms develop quite a strong resistance to these diseases over a period, and this prevents losses from becoming too serious, provided no unacclimatised sheep are introduced. On some hills lamb dysentery is endemic and the ewes have to receive regular vaccinations.

The main problems of hill sheep, however, arise from malnutrition but the hill farm improvements introduced to mitigate these problems are not all unhindered gains.

On many hill farms pulpy kidney disease of lambs and enterotoxaemia in ewes pose few threats and often no precautions are taken against them. Where an improvement scheme is introduced which increases the quantity and quality of spring grazing in a significant manner, the sheep should be vaccinated as losses due to the above diseases under such conditions can prove quite serious.

The simplest preventative is to give a combined vaccine which protects against a number of the clostridial diseases. Amongst this group of diseases is *braxy* which often causes trouble in hill hoggets, and *lamb dysentery*.

A very dissimilar group of complaints which can arise from fertiliser treatments are mineral deficiencies. Liming in particular can render such metals as copper and cobalt unavailable, or the availability is restricted to such an extent as to give rise in the case of copper deficiency to *sway back* in the lambs and, in the case of cobalt, to *pine* in the whole flock. Because of these dangers, a close watch should be kept on all stock running on newly improved land.

In many areas of Britain the hills are infested with ticks, as is the case in many other areas of the world. As ticks are major vectors of diseases, various diseases tend to be endemic amongst sheep flocks. The constraint that this puts on farmers in tick areas is that the introduction of new 'blood' can only be made via the rams. The introduction of ewe stock from a clean area usually meets with disaster.

SHEARING

On a hill farm shearing assumes much more importance than it does on the majority of low ground farms. It occurs later in the year than low ground shearing – usually late July to early August – and engenders by far the biggest concentrated effort that has to be made on a sheep farm. Childhood memories of a shearing in mid-Wales during World War I bring back a picture of a small child crouched over a pot of bubbling pitch, surrounded by voices of men, women, children, dogs and, above all, sheep. This is a vision of what was once a great social occasion in the hill lands of Britain. While the significance of shearing as a social occasion may have gone, some of the shearing is still done by 'neighbouring' and even on the hill farms the shearing gang has appeared. It is at this ewe clipping that the tallies are made and the hill farmer begins to put figures to his hopes and fears for the year's work. The store lamb and draft ewe sales which occur in late summer and early autumn will give the final firm figures.

WEANING

The ewes require at least two full months to get back into fit condition for tupping. This and the dates of the cost ewe sales set the framework for operations. Different parts of the world have different systems of disposal and those common to Scotland will be taken as an example.

What a lamb is like at weaning depends on when it was born, whether or not it was a twin, what its mother's milk yield was, its freedom or otherwise from disease or parasites and so on. The net result is that lambs weaned all at the one time show a great diversity of size and condition. The result of this is that lambs have to be divided up into categories. All the male lambs except a few retained

for breeding will have been castrated. These wether lambs are sometimes kept at home for feeding but, as the majority of hill farms have
no rotational ground or not enough on which to feed them, they
are therefore sent to the store sales.

Wether lambs are normally divided into four groups:

- *Tops*, or first quality wether lambs go for short keep and quick re-
  sale. At the present time, with a demand for lambs which carry
  very little fat, some lambs are sold straight from the hill farms
  to the slaughterhouse. In other cases, wholesale butchers attend
  the store sales and take out the best pens for immediate slaughter,
  feeding off on rape those that require further keep.
- *Seconds* need longer keep and are usually finished off on rape,
  clover aftermath, sugar beet tops, kale and the like.
- *Thirds* are animals for longer keep, usually fattened over the
  winter and early spring on kale, turnips, swedes, etc. South
  Country Cheviot wethers and similar slow-maturing, small sheep
  are ideal for this trade. Some may be carried through into the late
  spring and sold off grass before the new season lamb comes in.
- *Shotts* are the flock refuse, late lambs, lame lambs, bad doers and
  so on. A buyer with a good eye for sheep can often obtain lambs
  which can be profitably fattened indoors or in courts from this
  group.

The best of the ewe crop is retained for replacements, the surplus
going to the autumn sales. These lambs are usually acquired by
farmers from the upland areas who rear them and in due course
they are crossed by a longwool ram – as are the cast ewes – to
produce meat lamb mothers.

We have spoken of cast ewe sales and stratification earlier in the
book and all that remains to be said about the major differences
between hill sheep and low ground sheep management is about
tupping.

TUPPING

Where tupping takes place on the open hill as opposed to an enclosure
such as the ffridd, it is not normal to keel or colour mark the briskets
of the rams. In consequence the shepherd needs to pay particular
attention to each tup to ensure that he is working and seeking out
the ewes. To this end the animal must be fit but not fat. If he is too
fat, travelling the mountain will soon exhaust him; if he is too lean,
his libido is liable to be overcome by his hunger. The second point

is that no matter how fit and keen a tup is he will not be able to cope with as many ewes as a ram working in an enclosure. It is most important to get ewes in lamb at an early date and a flockmaster is well advised to have too many tups rather than too few, to the extent of always having a few in reserve.

The time at which hill tups are turned out is variable but it is naturally much later than for low ground sheep. A popular date in much of Scotland is 22nd November, the animals being returned to the steading on 31st December. This enables the shepherd to give thought to other matters on New Year's Day! This example is about the latest date for Britain as it applies to the high hills and mountains of Scotland where the ewes lamb on the hills. At lower latitudes such as the Welsh mountains or Dartmoor and in areas where the sheep are lambed on enclosed land and returned to the hills in the late spring, the tupping can be much earlier e.g. early November or even late October.

Where hill land is improved, e.g. the two pasture system, the time will arrive when it becomes necessary for the tups to be keel-marked in order that supplementary feeding prior to lambing can be controlled better and also that appropriate groups of ewes can be separated at lambing time to give them closer attention during lambing.

The next chapter will look at what happens to the ewes and lambs which leave the hills for lower and more fertile pastures.

Points to remember

1. Characteristics of hill farms in Britain and arid areas in other countries:

    Sour, peaty soils lacking in fertility and low in essential minerals especially calcium and phosphorus.
    Late-starting, slow-growing grasses, herbs and shrubs.
    In winter, wet, cold, snowy winters with little sunshine mean no growth of herbage.
    Growing season curtailed by extremes of temperature (or drought in arid areas).

2. Major problems:

    (a) For hill farms: winter feeding.
    (b) For arid areas: feeding and watering during drought.

3. Dominant vegetation:

    On hill farms: heather; mat grass; shrubs such as bilberry.
    In arid areas: stunted shrubs such as *Atroplex* spp.

4. (a)   Major weaknesses of sheep production on hill farms and
          in arid areas:

    Low lambing percentages.
    High neo-natal lamb losses.
    Poor lamb growth due to limited nutrition.
    Tick-borne diseases.

    (b)   Economic improvement depends on:

    Systematic amelioration of the most responsive land ('Improve
    the best first').
    Efficient control of stock.

5. Means of improvement:

    Reseeding.
    Fertilising.
    Drainage.
    Fencing.
    Mixed grazing.
    Irrigation where possible in arid situations.

6. Tupping

    Ensure that tups are fit but not fat.
    Tups must not be given too many ewes.
    Watch closely to ensure that they are working.

12 Further systems of sheep husbandry

Intermediate systems

The two large and obvious systems of sheep production, namely those of hill and semi-arid areas on the one hand and intensive meat lamb production from rich lowland pastures on the other, have sandwiched between them a series of very different methods of making a livelihood from sheep.

FARMING THE FOOTHILLS

In Britain there are large areas of the foothills where the land is enclosed, the major part of which is under grass. These are often referred to as the *rearing areas*, the land being considered not good enough for fattening cattle or sheep, and where the major form of production was traditionally the raising of dairy heifers, store cattle and store or breeding sheep. In modern times, however, with consumer tastes dictating a much leaner carcase in both sheep and cattle, these farms now produce a number of finished animals also. The majority of cast ewes from hill flocks find their way to these areas where they are crossed with longwool rams such as Border and Bluefaced Leicester, Teeswater and similar sheep. The ewe hoggets from the crosses may be sold on to other farmers for rearing or retained and overwintered to be sold as gimmers to lowland meat lamb producers. Likewise the wether lambs may be sold or fattened off rape at home.

In the past the major problem of these farms in the foothills was that of maintaining a high enough stocking rate to provide a good living while also avoiding serious health problems in the stock. There was a tendency to overstock some fields and understock others. There was also the habit of using specific fields year after year for the same purpose, to the extent that on many upland farms there was a field known as the 'lambing field'. This meant that the land became heavily infested with worms and bacteria. For many years stock on these upland farms suffered regularly during the winter from undernutrition due to the failure of the previous season's hay harvest. The increased popularity of silage making has helped to ameliorate this problem.

Space does not permit us to go into the details of every aspect of the management of sheep on rearing farms, but if good nutrition and hygiene are made first priorities the sheep husbandry should prove successful. If the flockmaster ensures that at no time does his flock suffer seriously from undernourishment or over-exposure to parasites or other disease organisms he will not meet with disaster. To this end, he will (a) plough to uproot sheep sick pastures, (b) provide fodder crops for supplementary feeding, and (c) he can also use other stock such as store cattle, thereby having a diluting effect on parasites. This *mixed stocking* also makes for a better utilisation of grassland than can be obtained from the use of one species only. The chief danger in this type of mixed farming is that, following initial success, the farmer can be tempted into over-stocking and finish up with an inferior product to sell. This in turn leads to dissatisfied customers and a resultant bad name. In the world of store stock production, a good name is most important. In addition to keeping a suitable stocking rate, the farmer needs to look to the mineral status of his soil and see that such activities as liming and slagging are carried out as appropriate.

A ROTATIONAL SYSTEM FOR UPLAND AREAS

A sound practical system of dealing with grazing problems of this type was evolved by workers at the East of Scotland College of Agriculture some years ago and has been adopted by many farmers.

Parasite control

The basis of the system is good parasite control, leading to much higher stocking rates than normal without a deterioration in the animals produced. The key point in the operation is that the young lambs are grazed on *clean pastures only*, the definition of a clean pasture being one which has not carried lambs or young sheep for at least twelve months. It is, of course, an advantage if the ground can be kept completely clear of sheep over that period. The point of keeping the land clear of lambs is that it is they who multiply the worm population in a spectacular way. Ewes are fairly resistant to helminths except at lambing time. If the ewes are dosed at strategic times, namely at tupping and just before lambing, then worm burdens remain low. If the lambs are subjected to a sound grazing regime they remain relatively free of infection as clean pastures can be achieved by a suitable rotational system.

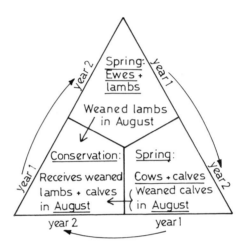

Fig. 12.1 Complete rotation system.

To take the example of an all-grass farm producing both sheep and cattle, the ideal is to be able to divide the grassland into three sections of equal size:

(a) The first third of clean grass goes to ewes and lambs in the spring.
(b) The second third is occupied by cows and calves.
(c) The remaining third is taken for conservation as either silage or hay.

In August the lambs which have not been sold are worm-drenched and weaned on to the conservation aftermath together with weaned calves and are left there until the autumn when the cattle are normally housed and the lambs are sold (see Fig. 12.1). When the cattle have been housed the ewes are drenched and allowed to graze until January on any area the farmer considers suitable. In cases where the ewes are wintered indoors they will normally be housed at the end of December.

From January onwards the ewes are confined to half the area they grazed during the previous spring and summer, the second half of the section being given the chance to recuperate and put forth some early spring growth. This last half of the area is grazed for a short period by the ewes and lambs immediately after lambing. The very young lambs suckling their mothers pick up very little infection. It is when they start eating large quantities of grass that heavy infection occurs if the herbage is badly contaminated. They are then moved on to the area that was occupied by cows and calves the previous spring and the following year they go to the land that was used for conservation in year one. On reasonably good land this sort of procedure should sustain a stocking rate of 16 or so ewes per hectare. This, of course, will depend on sound fertiliser treatment and proper timing of stock movements. The cattle equivalent of ewes is taken as about 5–6 ewes = 1 cow, but this, of course, depends on the size of the cattle and the sheep.

The success of the operation depends on two factors. The first is that the majority of common stomach worms' eggs and larvae do not survive much more than a year on the pasture, with the exception of nematodirus, so if a break of that length can be arranged there will be little build-up of parasites. The second is that cross infection between different species of animals is exceptional. The internal parasites which attack cattle do not usually attack sheep.

This system provides a sufficient time gap to prevent a build-up of parasites and enables both lambs and calves to move onto land which is relatively parasite-free at weaning.

It will be obvious that there are many upland grass farms where all the land is not suitable for conservation because of steep slopes or very broken ground. In such cases the 'three field' system described above will need to be modified. The procedure here is to make the conservation where mowing is practicable and alternate the sheep and cattle on the other two sections of the farm on a yearly basis. The ewes may be wintered on the conservation section of the farm or part of it (see Fig. 12.2).

A similar state of affairs arises on farms which have some arable land. The ewes can make use of stubbles, the lambs can be fed on rape or other succulent crop as can the ewes.

It is well appreciated that not each and every farm can be divided neatly into thirds or other fractions, but that is the sort of situation at which the farmer should aim. In all cases he should arrange the fields in blocks as far as possible in order to minimise stock moving problems. The guide the flockmaster has to follow are the points made in Chapter 3 on diseases and parasites, namely that the severity of a disease or parasite attack depends on how massive the infection is, the degree of immunity present and the nutritional status of the

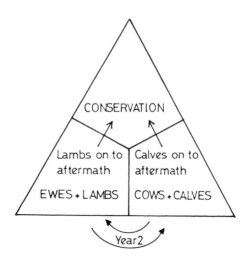

Fig. 12.2 Partial rotation system.

host. For instance, a lamb has low immunity to helminths, so the shepherd should keep these away from it, give it a good diet and it will prosper. Similarly, ewes need feeding well at tupping time and for a short time afterwards, again at lambing and during lactation and they too will perform well. The farmer's job is to fit his farm and his livestock together in such a way that these objectives are met and his farming should prove successful.

Hogget fattening outdoors

While the majority of lambs bred on lowland pastures and some from the good hills go for immediate slaughter, there are many hill lambs and some from the low ground which are fattened after weaning. Weaned lambs for further feeding before slaughter are known as *store lambs* (or *feeder lambs*) and should be selected with care, both in respect of the lambs themselves and their ultimate disposal. They should be drawn for age, size and weight. This uniformity minimises bullying at the feeding troughs and also makes disposal easier as the majority of sheep will be ready for disposal at the same time. This also makes it easier to judge how many animals will be required for the area of crop to be consumed.

In Britain there are two major groups of store lambs, wether lambs from the hills and foothills for the small hogget trade and Down cross lambs which give larger carcases, many of which go to the catering trade. The main crossing ram for this trade is the Suffolk, as it can be fed to higher weights than other crosses without becoming too fat.

In selecting the individual pens for feeding, the following points should be noted. Each lamb should have a bright eye with the head carried high and an alert look. There should be a good bloom on the wool. An inert, chalky-white fleeced sheep should be avoided. Signs of lameness and scouring should be absent. The farmer should learn to differentiate between lambs which are small due to parasitic infection and those whose lack of growth is due to limited nutrition. When buying sheep for further feeding one needs to know the type of ground from which they came as sheep, like other growing stock, should move from poor ground to better. This question of suitability of stock for a particular farm is such that many farmers obtain sheep from the same farm year after year.

A warning should be given against going for the very best lots in a store lamb sale. This is for a variety of reasons. One is that there will be wholesale butchers bidding and they can afford to pay more than the competing farmers. The old adage must be remembered that you have to keep even a well-bought bunch of stores a fortnight before they are worth what you paid for them. This is because each time you move sheep into and out of a market, or for any other reason gather them, they suffer some check in growth. A further reason for avoiding the apparent supersheep is that someone else may have attempted to finish them and, through running out of keep or for some other reason, has failed in his objective. Such sheep are always difficult to restart on the finishing process and by the time they are suitably finished may well have grown too big for their original purpose.

Mention has been made in Chapter 2 of crops grown specifically for sheep feeding. Which crop is grown varies from country to country and area to area. In high rainfall areas such as Britain and New Zealand cruciferous crops such as rape, kale, turnips and swedes are grown. In drier areas legumes such as lucerne (alfalfa), sainfoin, clovers, lupins and the like are favoured. In continental countries such as Australia and Canada, a whole series of crops will be used according to local climatic conditions.

How many sheep a particular crop will sustain and for how long is a matter for judgement on the part of the farmer. This is a problem which experience helps to solve, but some people have an innate sense which helps them to assess the potential of animals better than their fellows can, and these are the people who make money out of what, in many times and places, is not a particularly profitable exercise. A certain minimum time will be required to finish a given number of animals, and the area required will depend on the condition of the sheep, the weight of the crop and the supplementary diet to be used, if any. Care should be taken in the estimation of crop weights, and while a farmer cannot be expected to run round with quadrats taking random weighed samples, he should walk the whole area. This is particularly important in areas where the soil type is not uniform and with unpredictable crops such as rape. A crop which can look substantial on the headland can prove the reverse well into the field. Care must be taken to ensure that there is enough food available to finish the animals, otherwise they may have to be re-sold late in the season as forward stores and quite a lot of money may be lost. It is usually more economical to plough in a few turnips than be left with a flock of hoggets and no food.

It is very difficult to translate these general ideas into actual figures but a medium-sized hogget will consume about 50 kg of swedes or green material in one week. A useful figure for sugar beet tops is that tops from a good crop will feed 250 hoggets for one week on one hectare. With this crop residue the feeder must remember to wilt the tops otherwise oxalic poisoning may result. Wilting allows the oxalic acid present in the tops to break down into innocuous substances. The swede crop in a good swede area should produce 50 tonnes plus per hectare and the normal feeding period for the sheep will be about twelve weeks or a little over, allowance being made for wastage.

When the sheep for feeding arrive on the farm they should be injected against pulpy kidney disease and wormed. They should be introduced to the crop gradually, grazing it for only an hour or so each day for the first week. This opportunity is taken to reiterate that all major changes in an animal's food should be gradual. They should be run back on to a fairly bare pasture, and here it is of advantage to feed a little hay, say 250 g per head per day.

Whether or not supplementary concentrates are fed will depend on the farmer's objective, when he wants the animals ready, at what time and at what weights. Animals fed on such crops as kale and rape or legumes like lucerne will require an energy supplement only, and cereals such as oats, barley or maize will suffice. On the other hand, animals feeding off crops low in protein, such as swedes, turnips or sugar beet tops, will require a protein supplement in their ration. The concentrate ration should be arranged so as to contain about 14% digestible crude protein plus 3% minerals. The weight of concentrate to be fed varies normally between 120 and 150 g per head per day. The rate of live weight gain in fattening hoggets also varies but it is rarely spectacular; about 1 kg per head per week is normal.

The chief health problems to which feeding sheep are prone are clostridial diseases such as pulpy kidney. Footrot can also cause problems, especially under wet, muddy conditions. In the case of wether lambs one may also get the occasional case of urinary calculi.

Indoor sheep feeding

INDOOR HOGGET FEEDING

There are many circumstances where a farmer has a number of hoggets which are not ready for slaughter but has no forage crop or pasture on which their feeding can be completed. Under these conditions the animals can be fed indoors. On the other hand, a farmer may see an opportunity of buying store hoggets cheaply and subjecting them to the same treatment.

Indoor hogget feeding can be carried out in the same way as bullocks are traditionally fed, i.e. run in covered yards and fed with hay, sliced roots and concentrates. High quality silage can also be used. The types of store animals used are similar to those used for folding off arable crops. The sort of weekly ration given to each animal is as follows: 50 kg swedes, 3 kg clover hay and 1.5 kg concentrate such as barley. If the animals are not being fed a legume hay, the concentrate will require the addition of a protein supplement such as soya bean meal sufficient to bring the digestible crude protein in the concentrate up to 14%. This would also be the case where mangels were used. Lucerne or sanfoin hay are particularly useful in feeding sheep and alleviate the need for a protein supplement. Animals receiving roots in the above quantities will drink little, if any, water.

'BARLEY LAMB'

Circumstances sometimes arise where cereal grains are cheap and hoggets can be acquired at a reasonable price. Under such conditions it may be profitable to put the animals on to a high concentrate diet.

Some years ago a system of feeding Friesian steers was evolved at the Rowatt Research Institute by Dr Preston and others, based on cereal feeding with a protein supplement and little or no roughage. As was to be expected, others tried out the system on lambs and both were found to be quite workable. The lambs for this treatment can be housed in various buildings as were described for ewes. They can be on a hard floor with bedding or on slats. Hoggets on slats need about 0.55 m² of area for each animal. The slats need to be about 0.70 m clear of the ground. This allows for a feeding period of ten to twelve weeks.

As abrupt changes in feeding should never be made, a gradual transference is accomplished in one of two ways. A milled ration containing a large proportion of fibre is prepared and fed to the sheep. This ration has the proportion of concentrate increased until, over a period of three weeks or so, the animals are on to a fully concentrate ration. The other method is to introduce the lambs to the concentrate ration gradually. The lambs are started on the concentrate ration while still at pasture. This is gradually worked up to a quantity of about 700 g per head per day, also over a period of about three weeks. When this rate has been achieved, the lambs are brought indoors and the concentrate fed *ad lib*, together with access to ample hay. The hay is fed to appetite for a week and then reduced to 250 g per day.

A ration of this type which has been fed successfully is as follows: undecorticated cotton cake 5 parts by weight; rolled oats 3; rolled barley 4; oat husks 6; molassine meal 1.5 with 3% mineral added. The mineral is composed of equal parts of ground chalk, steamed bone flour and common salt. Synthetic Vitamins A and D are also added. Hoggets can be fed equally well on the ration normally fed to barley beef cattle, namely 8½ parts of rolled barley to 1½ parts of a commercial protein supplement, mineral and vitamin mix which is normally prepared in pellet form. When maize is cheap relative to barley, 2½ parts of barley can be replaced with cracked maize. In countries such as North America and quite larger areas of Europe, indeed anywhere where the conditions are such that maize is readily grown and where winters are severe, the fattening of housed sheep can be successfully carried out with whole maize cob silage. The silage is augmented with mineralised soya bean meal as in the case of swine. Fattening lambs will eat this ration avidly and make similar gains to those fed on what have become traditional barley mixtures. It should be underlined that for this rapid fattening the maize silage is whole cob, not whole crop.

The normal energy protein ratios for these rations is about six to one. Live weight gains under the system can be quite high, good lambs putting on as much as 1.5 to 2.0 kg per week. Food conversion ratios are about five or six to one.

The success of such a feeding regime depends on having a satisfactory price for the finished product. In other words, a lamb price as opposed to mutton. In addition there must be a cheap concentrate available and the farmer must have strict control over disease, and hence few losses.

Finally, the animals must pass through the system in a reasonable time. If not, too much food will be used for a profit to be shown. The usual cause of such problems is the wrong sort of sheep or the wrong sort of diet. Not only does this sort of situation put up costs, but the final product can be inferior and command a lower price than rapidly grown animals, hence compounding the farmer's problems.

Animals fed under the indoor regimes we have discussed are liable to the complaints common to all housed animals and, again, the necessity for good ventilation is emphasised. Sheep do not suffer from still cold, at least not unless the temperatures get very low, and provided the drinking water is kept liquid, the outdoor and indoor temperatures can be the same. In areas where winter temperatures are severe, heating the drinking troughs is necessary.

Of the specific problems which occur under a heavy concentrate feeding programme, bloat is the most important. This problem is normally prevented by introducing the animals gradually to their concentrates and then by ensuring that the concentrate is constantly before the animal, giving 24-hour access.

If the animals run out of food for a period they are liable to gorge themselves when the food reappears and bloat is the result. Continuous food is normally made available from hoppers, and care should be taken to avoid food 'bridging' in the hoppers as this, rather than the hopper being empty, causes the trouble. Any animal that suffers from bloat more than once should be removed from the system, put on to a less intensive regime or slaughtered. Acidosis — a metabolic disease associated with faulty digestion and similar to that occurring in 'barley beeves' — causes the occasional problem. Such animals should also go on to a less intensive system.

Urinary calculi cause more problems in wether lambs under intensive concentrate systems than under other regimes. The calculi block the urogenital tract, causing pain and difficulty in urinating. Such animals should be slaughtered as there is no economical treatment.

This intensive concentrate feeding system may not commend itself to the majority of sheep men. Apart from the associated digestive problems, it means in many countries that the lambs are competing with swine for concentrates although pigs are overall more efficient in the utilisation of concentrates than sheep.

There are, however, a number of situations where it could prove useful. On hill and upland farms when the autumn store lamb market has been weak, a number of lambs could be retained for fattening for the Christmas market. This could enhance the use made of a sheep or hogget house.

The use of housing in this manner on hill farms where an overall improvement scheme was in operation could prove most useful. It has been mentioned previously that twins are not normally desired on a hill farm, because if the ewe and her twins have to run on the hill the eventual outcome will be a very lean ewe with a pair of ill-thriven lambs. On the other hand, land for running ewes and twins is often very scarce.

An alternative is to split the twins at an early date and wean one, say at about six weeks of age. This lamb can be reared indoors either partially or entirely. Another method of solving the problem is to set aside a freshly reseeded small paddock where the early weaned lambs can receive supplementary concentrate feeding. The paddock can of course be just a small section of a larger reseed, fenced off with electrified sheep netting solely for the use of lambs.

When splitting mixed pairs it is obvious that the ram lambs should be removed so that the ewe lambs can join their peers from which flock replacements are ultimately selected. The farmer must make the choice between pairs of the same sex. He may feel that as the better lamb is thriving on the ewe he should let well alone and let the ewe keep the bigger lamb, hoping that he will rear a moderately good lamb artificially. On the other hand, he has to bear in mind that the larger, stronger animal is better able to fend for itself and will eat more concentrates and eat sooner than its smaller twin. This being the case, by letting the smaller lamb go to the ewe he could finish with two really good lambs.

Likewise on lowland grass farms, lambs which are not finished off grass could be subjected to similar treatment, while on an arable farm that fattens lambs on roots or sugar beet tops, late lambs could be brought indoors for finishing.

Finally, there is the situation in countries where, for religious reasons, the pig is not acceptable in which case the intensive indoor rearing of lambs could prove a most attractive enterprise.

The feedlot system

In many areas of the world where sheep are kept under range conditions and where rainfall is often limited herbage growth is insufficient to allow the lambs to come to slaughter weight in their first season. These lambs are moved on weaning to pastures in a higher rainfall area or to feedlots where they are fed on arable forage crops and cereals. Such feedlots are common in the western United States. Similar feedlots could be used in Britain for hill lambs.

There are various ways in which such a system could be organised. It could be a small operation run by the farmer, a large one into which he bought in lambs, or even a big co-operatively run business.

Housing for such an enterprise need not prove particularly expensive. The construction can be of cheap material, as previously suggested in Chapter 10. It is essential to have a roof that keeps the rain and snow out of the pens in high latitudes and provides shade in warmer climes. Providing shade is important as any animal that suffers from heat stress immediately responds with a reduced food intake and, in consequence, a slowing down in growth rate. The same is true of a water shortage. This is particularly so when animals are on a dry ration such as hay and concentrates. The most important requirement, then, in a hot climate is to see that the animal is neither too hot nor too thirsty, so that in addition to a satisfactory roof the building needs a continuous supply of good water.

Pens of an appropriate size, together with a supply of troughs and/or racks, complete the requirements. In a dry climate the floor is of little consequence and beaten earth is satisfactory. In a climate such as Britain, a raised floor to prevent water standing should rain blow in is an advantage. The floor can be of various materials such as brick rubble, rammed chalk, etc., but crushed clinker should not be used.

With regard to feeding in a situation where the most rapid liveweight gains are hoped for, the regime described under 'Barley Lamb' is suitable.

At this stage the point can be made that in areas where maize is grown, full crop maize silage can form an excellent food for sheep, both ewes and lambs. The author was very successful in substituting maize ear silage for barley when feeding 'barley lambs'.

Regarding whole crop silage, the Ohio Agricultural Research and Development Centre met with complete success in their investigation of corn (maize) silage over a period of years. They found it as satisfactory a diet as conventional hay and concentrates, provided the silage was adequately supplemented. The silage fed was fine-chopped, i.e. 125 mm in length with a dry matter content of 33–36%. The supplementation advised by the Centre was 1% urea, 0.5% ground limestone, 0.2% dicalcium phosphate and 0.05% sulphur. Most people appreciate the importance of a nitrogen or a protein source in the supplementation of maize silage as it is so much lower in protein than is grass silage, but the addition of sulphur to the ration is also of paramount importance. Silage of this kind has an additional point in its favour: it lends itself to mechanised feeding in a way no other foodstuff does. In southern and eastern England where heavy crops of maize for silage can be grown this can provide a simple system for feeding sheep over the winter.

Unconventional systems

In the normal pattern of sheep production ewes are mated in the autumn to lamb in the spring, their progeny intended for slaughter and fattened on grass or forage crops. There is no reason to stick slavishly to this pattern if the farmer sees an opportunity to earn money from an atypical system.

There is not enough space to comment on every sort of system but it may be helpful to describe a few unconventional ones to demonstrate how some farmers have been able to use sheep to exploit a farming situation which would otherwise have lain fallow.

WINTER LAMB PRODUCTION

The Dorset Horn breed of sheep have the characteristic, shared with the Merino and a few other breeds, of taking the ram at almost any period of the year. In addition, they are a breed of relatively high fecundity, are good milkers and with good fleshing characteristics. The ewes can be tupped (usually with a Down ram) to lamb in October and, with concentrate feeding, lambs are ready for Christmas and the New Year trade when there is often a strong market. Such flocks are kept on grassland dairy farms, the feeding patterns being such that the ewes do not clash with the dairy cows, the ewes being dry in the spring when the cows require the best and earliest grass.

Dorset Horn and cross bred ewes such as the Suffolk X Scottish Halfbred which take the ram early in the year can be used to produce lambs for the Easter trade. The ewes are put to the tup in August to lamb down in early January. This system also calls for winter housing and a considerable expenditure on concentrates. The ewes, as well as lambs, will require concentrate feeding during lactation; the amount fed to each ewe will depend on her litter size and can amount to 1 kg per day per head or even more. The lambs also require creep feeding. The feeding of roots or tubers to ewes under such systems is advantageous in that the appetite of the ewe is stimulated by the succulents and this is carried through to the milk yield.

A system such as the above can be practised with advantage on a farm that has suitable buildings and some rough, uncultivatable land where the dry ewes can be consigned away from stock that are in profit. They must, of course, be returned to good grass or a green crop for flushing.

ALL-THE-YEAR-ROUND LAMB PRODUCTION

The Dorset Horn with its extended breeding season can also be used in early weaning intensive systems and, theoretically, it is possible to get two crops of lambs per year. With a gestation period of about 5 months, however, this means very early weaning and little latitude for mistakes. Consequently, the majority of people who set off to exploit this all-the-year-round breeding propensity of this sheep normally aim at three crops in two years. The fact that such a system does not synchronise with grass growth tends to make it difficult to operate and, of necessity, it becomes very artificial. The fact that conception rates are not up to the average expected from conventional tupping times results in rather a large number of barren ewes at some periods of the year. These drawbacks are responsible for the system not achieving great popularity, although a number of people run such operations on quite a big scale.

Many indoor systems of sheep husbandry use early weaning. Also, in systems where the ewes often have large litters the availability of an early weaning facility is a great help. Sets of triplets and quadruplets can be broken up and where all the lambs are directed to a specific outlet, such as the Easter trade, this ensure that most lambs reach their objective.

THE EARLY WEANING OF LAMBS

As with all other new-born mammals, the lamb should receive colostrum in adequate quantity. If the lamb does not suckle the mother, colostrum should be administered by stomach tube as previously described. The next important point to remember is that ewe's milk is much more concentrated than is cow's, although Dr Ross at Stock successfully reared numerous lambs on 'neat' Jersey milk. Owen and Davis at Cambridge used a milk replacer successfully which compared with cow's milk as follows:

	Cow	Ewe	Replacer
Solids not fat	8.7%	11.3%	12.1%
Fat	3.6%	6.2%	5.3%

There are a number of commercial ewe milk replacers on the market and one of these should be used. They are adapted to the cafeteria system, i.e. the solids do not settle out but stay in suspension in the container. Lambs can be fed from a bottle, taught to drink from bowls, or fed from a cafeteria, a container from which a number of tubes lead off, each terminating in a teat, allowing lambs to be raised in groups. Cafeterias should be run and maintained strictly in accordance with the makers' instructions, hygiene being of paramount importance.

The simplest method with early weaned lambs is to leave them on their mothers for 48 hours. They are then individually penned and fed milk replacer from buckets, each of which has a teat. After ten to fourteen days, the lambs are bulked up and fed from a cafeteria. Concentrates are offered from ten days of age and this concentrate should be low in copper. It is best fed in pelleted form, small pellets to start with, i.e. through a 5 mm die, changing to a larger 10 mm pellet as they get older. The lambs are fed *ad lib.* until they reach 10–12 kg in weight and then weaned off liquid feed over a period of one week.

Where the lambs are trough-fed with liquid they are best housed in separate pens for the first 10 to 14 days. They should receive four feeds per day of about 300 g liquid per feed. Overfeeding with liquid is to be discouraged as the more liquid an animal receives above the optimum the less dry food will it eat. Hay and water should be offered from the fourth day.

After weaning the lambs can be transferred over a period to a home mixed concentrate and this can be based on barley and soya bean as previously described. From this stage forward, the treatment of the early weaned animal is as other indoor-fed animals.

The problems of these animals are as for other housed sheep — bad ventilation which can lead to pneumonia, insufficient floor and trough space which can cause nutritional troubles, and wet bedding leading to foot rot.

Copper poisoning also is always a threat. Copper pipes should never be used in the water supply as it takes very little copper to have an effect. *Pasteurella* infection and *orf* have given rise to serious situations on some farms, as has *acidosis*.

The main points the farmer has to watch are ventilation, hygiene and foodstuffs. If he keeps these under control his troubles should not be too great, but if there is a breakdown in health on any scale he should seek veterinary advice immediately. A disease outbreak in enclosed conditions usually spreads much more rapidly than outdoors with more serious consequences.

RAM BREEDING FLOCKS

It could be argued that there is nothing unconventional in ram breeding and that this heading is misleading. This is quite a valid criticism but in many areas of the world and in Britain in particular ram breeding tends to be specialised. This is especially so in the case of crossing tups such as the Down and Longwool breeds. It is also the case in some large groups such as the Scottish Blackface. It is for this reason that ram breeding has been allocated to this chapter.

Over the ages it would probably become apparent to herdsmen that small closed breeding populations tended to suffer a diminution in size and vigour over a period of time. On the other hand, in view of the fact that primitive hunter gatherer groups avoided this problem by tribal raids whereby they knocked the men on the head and carried off the maidens, they may have sought outside 'blood' from earliest times.

This decrease in physique is known as inbreeding degeneration, and it is countered by the introduction of genetic material from outside at regular intervals. In addition to the deterioration with inbreeding a number of serious problems can arise before general deterioration sets in where there are undesirable recessive genes present in the population. A common genetic abnormality of sheep of this type is that associated with the undershot jaw.

The earliest herdsmen doubtless started by capturing males from the wild, then relying on exchanging males with neighbours. This method still obtains in many areas. Finally, the provision of sires becomes a specialised enterprise. It should be noted that in the majority of cases the ram producers are also commercial sheep farmers. In some areas and amongst some breeds the breeding system tends to become hierarchical, with breeders selling to multipliers who sell to commercial users.

The main problem which can arise from specialised ram breeding is that a divergence of objectives can arise between breeders and users. Also the environmental background of the young ram may differ markedly from that in which his progeny will have to live. In the case of hill sheep the above problems can be quite substantial as it can in others of difficult environment. In these hard environments such as the hill farm there is the additional problem that while the hill farmer likes his wether lambs to mature quickly and flesh well, this requirement can be in conflict with the main requirement of the flock, namely the liveability of the females. The aesthetic appreciation of ovine excellence can also lead purchasers astray, as although good may be beautiful, beauty does not necessarily indicate commercial genetic merit.

The crux of the situation is that neither ram breeders nor users should pay undue attention to show points, and pedigree breeders should not become too distanced from commercial producers. While these strictures apply particularly to sheep kept in harsh conditions, they must not be overlooked in the breeding of sires of meat lamb mothers. For instance, certain rams may beget offspring that are lacking in initial vigour and not prepared to struggle for a teat. This could cause serious husbandry problems in meat lamb flocks where large litters are essential for economic success.

As was indicated in the reference to stratification, the cast ewes are served by crossing tups normally of the longwool breeds such as the Border Leicester or others of Leicester derivation, to give added size, prolificacy and milk production. These pure-bred longwools are kept in small flocks usually on mixed farms and usually on grassland. The regime under which they live is similar to that for any other breeding flock except that a great deal more individual attention is given to both ewes and rams. Attention at lambing time is particularly close.

The ultimate cross in the meat lamb producing system is usually a Down sire. These Down breeds are all descended from sheep that were folded on arable crops for a greater part of their lives. Today those arable flocks are in the main pedigree ram breeding flocks, and the flock spends a major portion of its time on crops grown specially for its sustenance.

Ram-breeding flocks on arable land
As has been explained, the Down breeds had their origins as arable flocks which served two purposes, one to provide good quality

mutton, the second to enhance the fertility of the soil. This meant that the breeder had to bear in mind ewes as well as rams, milk as well as meat, and also a reasonable prolificacy which ensured that the flock was regarded in a more all-embracing view than has been the case in the more recent past. The old arable system of general sheep farming foundered on the rocks of high costs and low returns. The modern tup breeding flock is assured of a reasonable return by its very nature, that is, of course, if it succeeds in producing good rams. The pedigree flock is therefore in a position to support higher costs than the more conventional enterprise. While pedigree sheep require more individual attention than commercial flocks, substantial improvements in labour utilisation have been made in recent times.

One of the major contributions to this saving has come from the use of the electric fence, both plain wire and netting. The handling of this equipment is much less energy-consuming than were the old wooden or metal hurdles. The two-strand electric fence can be used to divide up crop areas by the simple expedient of differential drilling. For instance, with a four section drill one box is filled with turnip or swede seed while the others contain kale. By this means bouts of six rows of kale and two of turnips are drilled. The electric fence is pitched along the turnip rows – the turnip crop being much shorter than the kale – so that the electric fence does not suffer from shorting. Paddocks can be constructed by means of electrified net. With a little ingenuity, creeps can be constructed where the nets join or by bridging over the creep in the case of plain wire. As with forward creep grazing, the passage ways through the wire are made big enough to let the lambs through but small enough to exclude the ewes. Creeping forward is important if the crop is being eaten clean off because, just as was explained in the case of forward creep grazing of grass, lambs must not be allowed to run short of milk in early life and of grazing later.

In addition to cruciferous crops legumes such as clovers, lucerne and sainfoin can be grown where soil conditions are suitable. On some farms residues of market garden crops are used. A substantial quantity of concentrates is also needed in such flocks for both ewes and lambs. As a large number of ram lambs are required to work in the tupping season following their birth, they need to be born early in the season, often under bad weather conditions, so ample feeding is needed to ensure that they are sufficiently well-grown to perform satisfactorily. In view of the bad weather normally attending their birth, sheltered lambing quarters are provided either in the form of a sheep house or a temporary fold with roofed pens and a cabin for the shepherd. Lambing is then carried out as in any other flock with the exception of particular attention being paid to identification and recording. This matter will be dealt with in the succeeding chapter.

A reminder must again be given that under any arable system where a lot of hay and concentrates are used, attention should be paid to the liquid water supply in case the supply of succulents falls off.

Finally, the point must be made that commercial ram production is not the job for the tyro. Rams are sold as much on the reputation of the breeder as on the apparent merits of the individual sheep, and it takes time and money to arrive at this position. Also, it requires skill on the part of the breeder or shepherd to present his animals for sale in an acceptable way. While the use of such cosmetics as bloom dip is to be deprecated, every effort should be made to present the ram in as tidy and attractive a manner as possible. It does not matter how high the genetic merit of a sire, if he does not look right he is unlikely to sell well. Those who have ambitions to excel as pedigree breeders need to serve an apprenticeship under a good shepherd with an established flock.

The reader, having had the main points of sheep husbandry brought to his attention, will probably speculate on what changes are likely to take place in the industry in the future and at what particular points progress is likely to be made. He will also want some indication of the most useful lines of approach in improving the performance of a particular flock under his management, and to these questions we turn in the final chapter.

Points to remember

1. First essential for high production from upland areas:

   Break the life cycle of internal parasites by judicious rotational grazing to achieve higher stocking rates.

2. For successful feeding of store lambs:

   Lambs should be drawn for uniformity in size, weight and age to

   (a) minimise bullying at the trough
   (b) make disposal easier since the majority become ready at the same time

(c)   make it easier to judge flock feed requirements etc.

Lambs should be free from obvious disease, parasites and lameness.

3.   Important requirements for success in intensive feeding of lambs:

Steady market.
Economical food source.
Good disease control.

4.   Major health hazards of intensive feeding:

Pneumonia.
Bloat.
Urinary calculi.
Pulpy kidney.
Foot rot.
Copper poisoning.
Respiratory diseases (therefore use well-ventilated buildings to promote good control).

5.   Barley lamb.

Start lambs on concentrates while still at pasture.
Make no abrupt changes in food, either in quality or quantity.
Basis of concentrates — barley or maize.
Before embarking on enterprise ensure suitable market paying high prices.

6.   Feedlot System.

Ensure protection from rain, snow (and direct sun in low latitudes).
Ample water and *regular* food supply vital.

7.   Winter lamb production.

Very expensive.
Housing needed.
Heavy feeding of concentrates for both ewes and lambs, hence farmer needs to be sure of a high price.

8.   All-the-year-round lamb production.

Two crops possible but very difficult.
Three crops in two years feasible.

Average litter size per pregnancy lower than for conventional lambing times.

9. Early weaning of lambs.

Adequate colostrum is vitally important.
Milk substitute must be properly constituted.
Cleanliness is essential.
If cafeterias are used, obey manufacturers' instructions.

10. Ram-breeding flocks.

Must lamb early — costs are therefore high.
Start with good stock.
Keep commercial excellence as the chief objective.
Do not overemphasise show points.
Present animals well at sales.
High prices are needed to offset high costs.
Serve an apprenticeship with a successful breeder.

13 Improving sheep production

On-the-farm improvements

Increased efficiency can be achieved on all farms, the good as well as the bad, provided the person in charge knows exactly what is happening and is prepared to make changes on the strength of the evidence. We must now look at the means whereby these farmers can identify and solve their problems.

The improvement of livestock production is too often taken only as the search for an outstanding sire which, once identified and extensively used, will transmute all to gold. While the importance of breeding cannot be disputed there are other aspects of improvement which demand attention, and a prompt solution to some of these problems may prove of more immediate help to the farmer than the introduction of a new breed or strain of ram. If a farmer wishes to make real progress in increasing the performance of his flock he needs a base line from which to measure progress and a method of monitoring his operations to see that objectives are being achieved.

While breeding and selection are not the only means of improving flock performance they must not be neglected. Selection must be carried out if for no other reason than to ensure that no deterioration in the sheep stock occurs. The majority of farmers already practise some selection — for instance, they discard broken mouthed ewes, those with mastitis, chronic lameness and barrenness. In the case of small flocks where the farmer knows each ewe individually this mass selection can be carried out quite satisfactorily. In the large, unrecorded flock, on the other hand, while the sick and the lame are just as readily identified, the poor performers may not be. Indeed, the eye-catching, well-fleshed ewe with the well-grown fleece could well be one which has run barren or suckled a single lamb indifferently. In an opposite case, a rather lean and tired-looking ewe may owe her poor condition to having nursed triplets well. In such cases, the selection of the apparently best on visual appraisal means going backwards rather than forwards. Adequate records must be kept in order to keep a check on all the parameters on which the running of a successful flock depends.

Record keeping

The records the flockmaster should keep depends on the sort of answers he seeks and the purposes to which he wishes these answers to be put. The records must be as simple as is consistent with acquiring the information desired. The keeping of records in themselves conveys no benefit on the sheep. All unused records do is to provide a chore for someone! It needs no great erudition for a person to be able to keep and interpret simple records but it does demand a good deal of self-discipline to see that the countings, weighings and bookings are done accurately and at the right time. It should be remembered that the more complex a system is the more likely it is that someone is going to let the process slip in times of overwork and stress.

The simplest record of importance to a sheep man is the number of lambs sold or retained in the flock compared with the number of ewes put to the ram. This figure can be compared with averages for similar breeds or crosses which can be obtained from organisations such as the Meat and Livestock Commission. These comparisons will tell him if his figures are good, bad or indifferent compared with those of other farmers but they will not tell him why. His figures may be bad because his lamb drop is low; this might be because his ewes were not adequately fed at tupping time or his perinatal deaths may have been high, and so on. It will soon be driven in on a farmer that if he wants any records in the least bit detailed he will need to be able to identify each individual animal.

IDENTIFICATION
There are two major methods of identifying sheep. These are by ear-tags or tattoos in the case of white-faced breeds. Ear notching can be used, as was popular with pigs, or branding with strong-horned breeds such as the Scottish Blackface. It is remarkable just how much information a horn can be made to carry by the judicial use of a branding iron. Unfortunately this method has to wait until the animal is of some age before it can be used.

Once a recording system has got under way lambs will need to be tagged at birth. The best method to adopt is to use chicken wing bands. These will suffice until the lamb goes for slaughter or is weaned. The weaned lamb can then receive a metal or plastic adult tag or be tattooed. The chicken wing bands should be purchased with duplicate numbers and one placed in each ear as, in common with all ear tags, some will get torn out. To make identification doubly sure blanks can be acquired and, with a set of appropriately sized punches numbered 0–9, further replacements can be made available. The reason for starting with wing bands is that they are very light: they do not drag down the lambs' ears and cause less ear damage than adult tags. The tags are attached by a sharp-pointed pin which readily pierces the ear and is clipped against the tag with a small pair of pliers. The adult tags can be either a single strip of metal which is bent round to clip against itself or the plastic press stud type in which two pieces of plastic clip together. Plastic has an advantage in that it can be in readily distinguishable colours; these can be used for group identification which can be most helpful when sorting sheep through a shedder (i.e. an arrangement of gates for separating sheep into groups). In sheep with white ears tattooing is an excellent method of identification but is unreliable where animals are of very dark or broken colour. The chief point to emphasise regarding tattooing is that the operator must ensure that he gets the job right first time. A second attempt at this operation usually results in a jumbled set of characters which are quite illegible.

The control of sheep

No useful purpose is served by being able to identify each sheep unless one knows what food it has eaten, with whom it has mated and so on. A farmer must, therefore, have proper control of his animals. He must also have a readily workable handling system whereby the sheep can be manipulated with a minimum use of labour. This means that the farm must be well-fenced and have a well thought out system of pens whereby weighing, injecting, foot rot and other treatments can be carried out.

Handling pens should be on rising ground if at all possible. The area should be well-drained and, where practicable, water and electricity should be laid on. In countries with variable and unpredictable weather such as Britain, it is a great advantage to have at least some of the handling facilities under cover. The ability to perform such work as routine weighing, inoculations and point scoring away from the elements is a great boon.

SHEEP HANDLING PENS

There are three basic types of handling pen all of which can prove satisfactory. We have already mentioned light tubular steel hurdles which can be transported round the farm and set up in a field any-where near the flock. These hurdles can be bought with feet which fold along the length of the hurdle for transportation. These feet are turned at right angles for erection and are quite suitable for setting up on concrete for indoor use as well as outdoor. There are manufacturers who produce complete systems in galvanised tubular steel. These can be permanent or semi-permanent and at their most elaborate contain not only pens, weighing and shedding races but also dippers and footbaths. The third arrangement is for the farmer to build a wooden set of permanent pens using farm labour or the services of the local carpenter.

Whichever system is adopted the essentials are as follows. A series of handling and holding pens connected to a shedding race which should also incorporate a weighing machine are all necessary. The pens next to the race should be forcing pens in order that the sheep can be funnelled into the race. It is also convenient to have a catch-ing pen where the arrangement of gates is such that an individual sheep can be caught without difficulty. This point of easy handling needs to be emphasised as the easier the sheep are to handle the less will they suffer from bruising and stress. At the same time there are occasions when the sheep need to be assessed manually. The weigh-ing machine is not able to differentiate between a large, lean sheep and a small, fat one. Figure 13.1 shows a general lay-out of such a series of pens.

While the sizes of the pens are in no way critical, those of the race are. The race must not be so narrow as to prove difficult for the sheep to negotiate, especially an in-lamb ewe. It must not be so wide that the sheep can readily turn round and try to work their way back against the flow of animals. The measurements will, of course, vary with the size of the sheep. For very large sheep, such as the Lincoln, they will need to be substantially larger than for a very small breed like the Welsh Mountain.

Race measurements which have been found suitable for medium-sized ewes are as follows: height 1.0 metres, width 0.40 metres. Some people favour sloping sides, in which case the top could be slightly wider, say 0.45 metres with the bottom 0.30 metres. The sides should be lined with smooth sheet metal from ground level up to 0.20 metres. This prevents sheep from getting their feet caught.

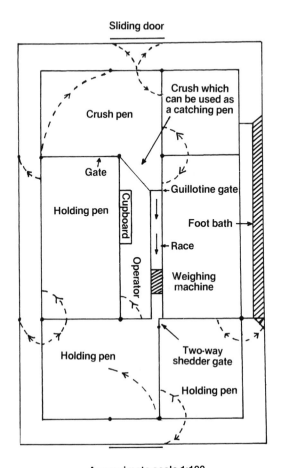

Approximate scale 1:100

Fig. 13.1 Sheep handling shed. (*Note*: the 1 m alleyway round the building increases the cost but the advantage of being able to recirculate the sheep is substantial.)

There should also be a ramp up to and down from the platform of the weighing machine. The weighing machine should be let in to the race just before the shedder. The length of race that needs to be removed to take the weighing machine crate is about 0.85 metres. The top of the weighing machine platform will be about 0.15 metres above floor level. In the line of approach of the sheep to the weighing machine a section of the top of the race adjacent to the operator should be hinged so that it can be dropped down. The size of this section should be about 0.75 m × 0.20 m. This enables the operator to get at the sheep and check its ear tag before it enters the crate. If the weighing machine is not needed elsewhere it can be left in the locked position to prevent damage and sheep can then be run over it. In situations where a number of weighings has to be made the animals learn to go through the exercise without hesitation. Figure 13.2 shows how a series of doors on gates can be arranged in a race to provide for two- or three-way drafting or shedding.

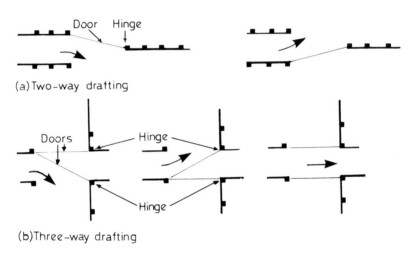

(a) Two-way drafting

(b) Three-way drafting

Fig. 13.2 Shedders for drafting sheep. (*Note*: the drafting doors are made to swing as shown.)

Having organised the identification and handling and weighing of the sheep the farmer is now in a position to keep the sort of records that will provide him with a great deal of physical data about the performance of his flock. This is, however, not all he needs to know. As stated previously, the objective of a progressive farmer is not to achieve the highest lambing percentage in the area, sell the highest priced lambs or to win the most awards at the local show. His objectives are, or should be, to have one of the most consistently profitable flocks while, at the same time, adding to the capital value of his flock and farm. He will, therefore, need to know what inputs he is using to achieve his physical results. He will need to keep grazing records, records of crops grown specifically for the sheep, costs of bought-in feed, veterinary services and of any other outgoings attributable to the sheep. In effect he will need to be able to make a gross margin analysis. He can compare these with those of other farmers and get some ideas of his financial efficiency. This, amongst other things, requires that the farmer makes an accurate valuation of the flock each year and ensures that as far as is practicable a proper age balance is kept.

Control of physical and financial resources

Although the chief concern of this book is animal husbandry, a few words need to be said about general farm management. If the sheep performance records are to mean anything the other farm records must be accurate. For instance, there needs to be a barn book, especially on a farm that has more than one livestock enterprise. One needs to know what foodstuffs and other consumables have come in, what they cost and to which enterprise they have been issued and charged. At an appropriate time of the year, usually early autumn, a computation must be made of the fodder on the farm and also an assessment of what the stock will require over the winter. If there is a shortfall, arrangements must be made to provide supplementary feed. It is no use waiting until a flock of sheep is storm-bound and starving before seeking extra food.

FENCING

There is no point in keeping records of where sheep are supposed to be and on what crops they are thought to be grazing – particularly in a pedigree flock where it is essential to be sure of where the rams are and with what ewes – unless fences are of appropriate construction for the job and kept in good repair. There is not enough space in this book to go into all the different sorts of fencing that can be used to control sheep but a little should be added to what has already been said about hill fencing. It is obvious that the sort of fence which is adequate to contain feeding hoggets on a root break in a country such as Britain or New Zealand is not the sort of fence to use for containing ewes and lambs in a country where coyotes or similar animals are a menace.

The problem of anti-predator fencing is now well under control. The *Piesse fence* (an Australian invention) has the live and earth wires alternating from top to bottom of the fence. There is also a live trip wire 0.20 m from the fence on the outside and 0.15 m above ground level. Previous electric fences where there were two or three live wires were not effective as the intruders were not normally grounded when they touched the wires and hence received no shock. This fence is reliable and less costly to erect than is conventional post and wire. In most countries information can be obtained about such fences from advisory services; the United States Department of Agriculture, for instance, produces a leaflet on the erection of such a fence.

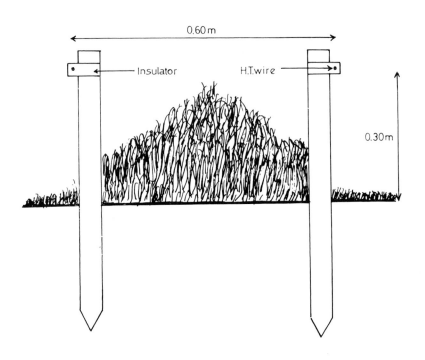

Fig. 13.3 New Zealand Grass Fence. In Australia hardwood battens (e.g. Jarrah) are now being used. This wood does not conduct electricity, hence additional insulation is unnecessary.

Much experimental work on low cost fencing has been carried out, largely in New Zealand. One fence of outstanding interest and now available in Britain is the *Grassfence*. This is a high voltage system where two live wires are run parallel in the horizontal plane, 0.60 m apart and 0.30 m from the ground (see Fig. 13.3). The grass grows up between the two wires and this forms a visual barrier which provides a psychological check to the animals before they even touch the wire. A probing of the fence is punished by a painful shock which reinforces the first suspicion that intimate contact with the hedge should be avoided. If greater security is desired a further energised wire can be introduced offset between the parallel wires and about 0.30 m to 0.45 m above them. This fence can also be used to render an old fence or dyke stockproof. All that is necessary is for a wire to be run on either side of the old fence. This should prove of great help in countries such as Canada and the northern United States of America where fences are heaved by winter frosts. Similarly, in areas such as highland Britain where many of the old traditional dykes are in poor repair and costly to renovate the electric fence could play its part. They can also be used as outriggers to old existing fences.

The secret of success when using electric fences is to have a really good earth contact.

Sheep improvement schemes

Having discussed the control and recording of sheep, attention must now be turned to how this information can be used to improve sheep performance. One way is to improve the animals' environment. Greater care, for example, can be taken at lambing time to increase the percentage of lambs reared; improved feeding of the ewes can also enhance the lambing percentage. Improved grassland husbandry can increase the amount of feed available, thus making a higher stocking rate possible, and so on. Increased efficiency in matters such as these will almost certainly lead to better performance but in some instances it will become apparent that there is only limited potential for improvement in the sheep stock.

Where the potential of the flock is considered unacceptable the farmer has two choices – either to employ another breed of sheep or to improve his flock by selection and the introduction of outside genetic material. Such genetic improvement is best achieved by seeking the help of someone experienced in such improvement programmes and by joining with other flockmasters of a like mind. In most developed countries he will be able to get the opinion of experts from county colleges, government advisory services and, in Britain, the M.L.C. to advise him on the best system for his purpose.

The first national scheme for sheep improvement was inaugurated in Finland in 1918. Their main objective was prolificacy in which they have met great success. The majority of developed countries have sheep improvement schemes promoted by government or states (as in New Zealand), wool marketing boards, breed societies and the like. These vary in the amount of information recorded and the closeness of supervision of co-operating farms by officers of the organisation. They may be comprehensive, as in New Zealand or the emphasis may be on breeding performance as in Britain, with wool being subsidiary, while in New Zealand and Australia wool is of great importance. In France and some other countries there are societies which record milk yield. In addition to recording, some organisations have established testing stations. The Ontario Ministry of Agriculture has two centres which operate recording schemes and, in addition, offer farmers the opportunity of sending in lambs for individual performance testing. Such facilities are of particular importance to breeders of terminal sires for meat lamb production. In the U.K., the M.L.C. have an individual ewe recording scheme for pedigree flock owners. Anyone wanting a comprehensive view of recording should consult *Performance Recording in Sheep* by J.B. Owen.

Schemes in use over various parts of the world differ in their particular requirements but the basic information normally required is as follows: date of birth, sex, weight, number in litter, sire and dam identification. It is also helpful to give details of birth, if assisted, etc. The lambs are normally weighed at fixed intervals and the Ontario scheme records the 50 day and 120 day adjusted weights. The M.L.C. takes lamb weighings at 8 and 12 weeks. In addition the M.L.C. requires the weight of the ewes at tupping. Finally fleece weights can be recorded. These records are computerised, the raw weights being corrected for litter size, sex and so on. It is from these figures that performances can be worked out for both lambs and ewes. If a breeder is eligible he is well advised to join in a scheme such as that described above as the stimulation of working with other breeders and the discipline of keeping records in accordance with the dictates of someone else usually lead to a more thorough application to the job.

The flockmaster may be interested in breed improvement within his own flock but realise by virtue of numbers that his progress is liable to be very slow and decide to join with others in a private improvement scheme.

CO-OPERATIVE BREEDING SCHEMES

These have become important in Australasia and have arisen from the fact that a number of knowledgeable breeders were aware that the important feature of improvement is selection pressure. In order that selection pressure can be increased the numbers of animals in a scheme need to be large. These schemes comprise a number of breeders who have got together and pooled their resources. A nucleus flock is formed from a selection of the best females from each flock. Sires from the nucleus go back to the home flocks. Figure 13.4 shows how the system works.

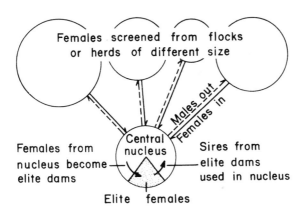

Fig. 13.4 Simplified structure of a cooperative breeding scheme.

The most important point in such a set-up is that each co-operator has complete confidence in his co-partners and that the nucleus flock is selected and managed by a manager who has complete authority and control. The criteria for selection should be clearly laid down and the emphasis should be on commercial traits and not aesthetic considerations. A decision has to be made on the ratio of ewes to the nucleus and tups out. A useful figure is four females in to one male out. This at least would do for a start — subsequent changes could be negotiated.

The main anxiety in such schemes is the spread of disease, especially of the slow virus type such as scrapie where the incubation period is very long. It is not difficult to envisage the position of one in such a group who was held to have introduced such a disease into the organisation. In spite of this difficulty a cooperative breeding scheme provides one of the few opportunities by which a farmer, other than a very large one, can be involved in a worthwhile and important programme.

USE AND INTERPRETATION OF RECORDS
Having said this, the breeder should keep records even if only private ones. What kind of records will depend not only on the predilections

of the farmer but the type of flock he runs. The ease with which recording can be carried out varies a great deal between a high hill flock and a closely shepherded intensive one. The pedigree man who is in an official scheme will probably have a fieldsman to keep him right so here are some suggestions for flocks that are not self-replacing.

As lambs have to be tagged at birth no great additional effort is required to weigh the lambs and record the weights. All that is required is a spring balance and a sling made of canvas, hessian or some similar material. Stillborn lambs should be weighed and recorded and a note made of difficult lambings. The deaths of ewes and the cause of death should also be recorded.

In meat lamb producing flocks the lambs need not be weighed again until near the time of slaughter. When the lambs go for slaughter their weights and dates of departure will be recorded together with the price received. When the point scores of the ewes at tupping have been added to these records the farmer has the basis of sound management control in hand.

The following are examples of the sorts of information which can be put to fairly immediate use in the improvement of flock performance.

- The first thing to be noted at the actual lambing is the number of *still births*. If these are excessive the veterinarian should be consulted.
- Similarly, *perinatal deaths* noted in the record book bring home the extent of this form of loss in a way memory does not.
- The *birth weights* of the lambs should also be examined and if the majority of dead lambs are found to be much below the average expected the feeding of the ewes has to be questioned.
- Another example is the question of *large lambs* giving rise to *difficult lambings*. Are they all from the one tup? Or, if not, have the ewes been over-fed?

At the end of the season an assessment is made to see if the majority of lambs met the market for which they were intended. If not, where did the fault lie? Did the ewes give too little milk, either because they were of low milk-producing potential, too low in condition at lambing, inadequately fed during lactation or were the sires of too low a growth potential for the object in question? There are many other aspects of sheep production on which records like these can throw light on suggested improvements. Records can also be useful in the long term in that if a flock runs into a health or fertility problem, it is most useful for a veterinary investigator to be able to look back over some years of recorded performance.

In addition, the farmer could well find the compilation and perusal of such records intellectually stimulating and conducive to a really keen approach to his farming which could prove not only interesting but profitable.

In pedigree and other self-replacing flocks it is also the role of the records to provide information which can be used in the genetic improvement of the sheep. The farmer who has no knowledge of genetics needs guidance from a geneticist or someone knowledgeable in the subject. For those who wish to refresh their minds, suggested titles are given at the end of the book.

The problem with livestock improvement is that there is a number of superficial features such as coat colour, presence or absence of horns and various other features that can be readily changed by breeding. The genetics of these conditions are simple, being determined by a few genes and, in consequence, their heritability is high. This has encouraged people to think that livestock improvement is much simpler than is the case.

Economically important characteristics tend to be multifactorial in origin and have low heritabilities, hence improvements usually take a number of generations e.g. longevity and milk yields. Heritabilities are expressed in figures from 0 to 1. 1 is the figure for characteristics which are 100% certain, such as having four legs. Examples of other figures for heritabilities are milk yield 0.10–0.20, rate of live weight gain in lambs 0.10–0.30, wool weight 0.30–0.45, fineness of wool 0.40–0.70. These last figures come as no surprise as the most outstanding improvement in sheep in the last century has been the improvement in quantity and quality of wool in the Australian Merino.

SELECTION

The most difficult problem in livestock improvement is selection. In most cases the farmer wishes to improve a number of qualities in his animals simultaneously. His first realisation if he tries to put such a policy into effect is that if the selection is rigorous he will soon run out of females. This is particularly so for high hill farms in a bad year. Almost all the ewe lambs need to be kept merely to ensure that ewe numbers are not allowed to fall.

To counter this the plan adopted may well be to go for one particular objective and disregard others until the first goal is at least in sight and then tackle the others. On the other hand, much slower progress towards the attainment of all his objectives may be the safer course for the breeder.

The above problem is a case in point where the advice of a sheep geneticist should be sought. Quantitative genetics is a matter of probabilities and what the livestock improver wants to know is the odds in favour of different approaches. There is also the possibility that some of the factors which the farmer has in mind to improve may have a negative correlation and, once again, the importance of professional advice is underlined. For example, one cannot help but be struck by the high quality and quantity of wool produced by the Australian Merino and New Zealand Romney and also be surprised by their poor litter size.

In view of what has been said about heritabilities and the limitations that can be made in selection on the female side it might be better to introduce genetic material from another breed rather than use the slow process of within breed selection. This indeed is what many are already doing to increase fecundity by introducing Finnish Landrace sires.

Stratification has been discussed earlier; this is a case of supplying such genes as those for superior milk production and growth rate by the shortest possible route. On the other hand, suppose it was felt desirable to alter the fleece of, say, the Scottish Blackface. This could be done fairly readily by within breed selection.

Before embarking on any breeding programme the flockmaster must make up his mind and be quite sure of his objectives because to chop and change can be expensive. He must also be prepared to abide by his own decisions and stay with them in whatever he sets out to do. If, for instance, he is in search of a minimum care flock he must not turn soft-hearted and say, for example, that a certain aspect of the situation in a bad year was atypical and the circumstances of the flock should be ameliorated.

Finally, anyone determined to press on to what he feels is the ultimate goal should beware the parochialism of time. That which looms large in the here and now may, in a few years, have receded into insignificance. In war-time Britain of the Forties the cry was: 'What are we going to do for fat'; today it is: 'What are we going to do with the fat'!

Future developments, official bodies and the farmer

Clearly, although there is much that can be achieved by the individual farmer, some programmes demand the resources of government bodies, marketing boards and research organisations. For example, the resolution of the problem of copper metabolism, copper poisoning and sway back disease far exceeds the competence of one scientist to solve, let alone a single farmer. It requires geneticists, biochemists, veterinarians and people from various other disciplines working in cooperation before a clear picture can be obtained.

An even more esoteric development which could well be of outstanding help to the industry is genetic engineering. If, for example, a number of genes from various species could be rearranged to enable nitrifying bacteria to live symbiotically with grasses such as agrostris as well as legumes such as lucerne, hill farm improvement would benefit enormously.

In the field of purely technical innovation the introduction of a cheap and simple means of determining multiple pregnancies at between, say, 80 and 100 days from conception would enable the feeding of the in-lamb ewe to be carried out with more precision than at present. Ultrasonic pregnancy detectors are already being developed which show promise for the future.

The national or, indeed, the international requirements for increasing meat production may well be in conflict with the immediate best interests of the individual producer. There is no doubt that one of the simplest ways of increasing a meat supply is to slaughter animals at a greater weight than is normally the case. In a world where the gap between food supply and mouths to be fed grows even larger it would seem that global strategy demanded heavier carcases. On the other hand today more affluent customers are demanding smaller, leaner carcases – at least, this is the case in Western Europe. It could well be that one way in which this problem could be resolved would be for producers and handlers to take a new look at butchering and retailing. The poultry industry's success in marketing turkey portions suggests that there could be a significant market for various lamb joints that have been boned out and rolled. This market could provide a steady outlet for 50 kg live-weight hoggets.

Government intervention is the only effective means of controlling such diseases as foot and mouth, sheep scab, blue tongue and scrapie. The most important action farmers can take is to run closed flocks wherever possible or replenish from a known and trusted source. If any such disease is discovered it should be reported to the proper authority.

The farmer may well ask what he can do to influence developments or government policy. The answer is that he should exercise his mind on these problems and let his views be known to the appropriate pressure groups. In Britain, for instance, the farmer can make his voice heard through such organisations as the National Farmers' Union. They, in turn, can represent to government the need to provide money for technical research, they can say what regulations they feel are necessary to protect their flocks from transmissible diseases and so on. It is also the duty of such organisations to impress on their members the importance of strict adherence to animal health regulations.

On the marketing side the farmer can join up with like-minded producers to form co-operatives because it is through such organisations that a regular supply of a standard product can be achieved. This standardisation should enable a firm long term market to be sustained.

In the field of breed improvement co-operative association with other breeders is virtually essential. This is in order to get large enough numbers of animals into a breed improvement plan and to permit the rigorous selection which is necessary if the plan is to succeed.

These then are the broad fields where authority must act but the cumulative effects of the actions of individual practical farmers can be far from negligible.

Points to remember

1. The need for records.

    The main essential of any improvement plan is a *base line* from which progress can be measured. Therefore *records* must be kept. Records must be:

    Accurate.
    Timely.
    As simple as is consistent with the information required.

    Records must be *financial* as well as *physical*. For example:

Amount and cost of homegrown and purchased food used.
Area of pasture allocated to sheep.
Cost of veterinary attention.
Medicines, dips and labour costs etc.

2. Physical details.

(a) Physical records for *extensive flocks* can be simple and on a group basis. For example:

Number of ewes put to the tup
Number of lambs reared
Number of ewe deaths
Number of ewes barren

(b) *Intensive* flocks require individual records. For example:

Which ewes to which tup
Date of lambing
Number and sex of lambs
Weight of each lamb
Still-births and lamb deaths
Number of dead and barren ewes
Weaning weights
Individual ewe wool weights at shearing, etc.

3. Individual records.

Accurate individual records require:

(a) Positive identification. Methods include:

Ear tags
Ear tattooing
Ear notching
Horn branding

(b) Complete control which also ensures controlled matings. Methods include:

Post and wire fences (keep wires taut!)
Electric fences (ensure good earthing!)
Sheep netting (plain or electrified)

(c) Adequate handling pens.

These are necessary to ensure positive identification and the efficient use of labour.

(d) Accurate weighing machines.

4. Analysis of records.

Analyse all records at the end of each season to assess the strengths and weaknesses of the system. For example:

Is the number of perinatal deaths high?
Are lambs slow in arriving at slaughter weights? etc.

5. Genetic improvement of the flock.

- Large numbers of sheep are needed to ensure success.
- Schemes for improvement are usually cooperative.
- The scheme needs a well thought out plan with limited objectives which must be strictly commercial.
- The scheme must be under the control of one man.
- He must be, or be advised by, a geneticist with substantial experience of large domestic animals.
- All cooperators must have full confidence in one another and in the manager.
- Great care must be taken to exclude disease, especially of the slow virus types.

Book list for further reading

It is hoped that this book will have stimulated readers to further their knowledge of sheep husbandry and the following books are recommended. A number of these books contain a wealth of references to original work which opens up the whole field of the science of sheep production to the student. The order in which the titles are arranged is not intended to convey any order of merit. It is merely to reflect their relationship to this volume.

Sheep Husbandry and Diseases (6th edition)
A. Fraser and J. T. Stamp revised by J. M. M. Cunningham (Granada)
Breeds – distribution – general – health – history.
(The latest edition is still in the course of preparation at the time of publication of this book.)

Sheep Production
J. B. Owen (Bailliere Tindall)
Sheep distribution – growth – reproduction – genetics – breeding.

Animal Nutrition
P. McDonald, R. G. Edwards and J. F. D. Greenhalge (Longmans)
A standard text book of nutrition.

Technical Bulletin 33: Energy Allowances and Feeding Systems for Ruminants (H.M.S.O.)
Rationing on the basis of metabolisable energy.

Improved Feeding of Cattle and Sheep
P. N. Wilson and T. D. A. Brigstocke (Granada)
Rationing of ruminants.

Feeds and Feeding
F. B. Morrison (Morrison Publishing Co, Iowa, USA)
An American standard text on animal nutrition.

Feeding the Ewe
Meat and Livestock Commission Technical Report No 2
Feeding regimes for breeding ewes.

The Shepherd's Guide
J. A. Watt. Department of Agriculture and Fisheries for Scotland
Advisory Bulletin No 9 (H.M.S.O.)
Sheep diseases.

Sheep Production
Andrew M. Speedy (Longmans)
General — upland grazing — disease control — handling equipment —
breeding.

Sheep Production and Grazing Management
C. R. W. Spedding (Bailliere Tindall and Cox)
General — pasture management — production — ecology — animal
behaviour.

Profitable Sheep Farming
M. McG. Cooper and R. J. Thomas (Farming Press)
General — grazing systems — health.

Intensive Sheep Management
Henry Fell (Farming Press)
General — breed distribution — indoor management — marketing.

Science and Hill Farming
Hill Farming Research Organisation 1954–1979 (H.F.R.O. Bush
Estate, Penicuik, Midlothian)
Hill farm improvement.

An Introduction to Practical Animal Breeding
Clive Dalton (Granada)
Basic genetics for breeders of farm animals.

Performance Recording in Sheep
J. B. Owen (Commonwealth Agricultural Bureau)
Recording systems.

Modern Aspects of Animal Production
N. T. M. Yeats (Butterworths)
Recent scientific advances and their efffects on animal production.

Glossary

Acidosis: an acid condition of the blood usually caused by over-feeding with cereal grains.

Aftermath: the regrowth of herbage after mowing.

Agisted: an animal taken on to land for grazing by one farmer on behalf of another, usually at a price of so much per head per week, is said to be agisted.

Antibody: the proteins produced in an animal when a foreign substance has gained access to it, e.g. a bacterium or virus. The antibodies of the host combine with the antigens of the invaders and neutralise them, hence giving protection to the animal.

Bloat: an abdominal swelling in a ruminant caused by gas being trapped inside the rumen.

Cast: (a) a ewe removed from the flock on account of age or infirmity (b) a sheep or other animal which has rolled on to its back and cannot regain its feet is said to be cast.

Colostrum: the milk-like liquid given by a mammal immediately after parturition.

Crutching: the removal of wool from the breech of a sheep to prevent soiling.

Dystocia: a malpresentation or difficulty at parturition.

Entire: a male animal which has not been castrated.

Ewe: a female sheep which has given birth.

Feeders: young sheep which require further feeding after weaning to make them ready for slaughter.

ffridd (Welsh): a lower and more sheltered enclosed area of a mountain grazing.

Finishing: the process of feeding animals to slaughter weight and condition.

Flushing: the rapid building up in condition of ewes by extra feeding prior to tupping.

Foggage: herbage growth which is left unmown for animals to consume *in situ*.

Folding: sheep which are closely confined on an arable crop such as swedes are said to be folded.

Genes: the basic units of inheritance.

Gimmer: a young female sheep, normally a maiden aged about 15 months plus.

Heft: an area of a hill sheep farm with which a particular group of sheep is associated.

Hefted: a group of hill sheep which have a particular affinity for an area of ground and from which they do not normally stray are said to be hefted.

Heritability: the strength of inheritance of a trait.

Heterosis: the vigour obtained by crossing two distinct lines of animals or plants.

Hirsel: an area of a hill sheep farm made up of a number of hefts. Normally bounded by natural features such as water courses, but not necessarily fenced. Usually the charge of one shepherd.

Hogg or hogget: a young sheep of either sex beyond the lamb stage.

Hypothermia: a condition of markedly sub-normal body temperature.

Immunity: the presence of antibodies against a certain disease in the blood of an animal in suitable quantity confers a protection against that disease. This is known as immunity. Immunity is of two types — active and passive.

Active immunity is where the animal has been induced to produce its own antibodies, either by contracting the disease, or by the introduction of killed or attenuated organisms into its body by vaccination.

Passive immunity is where the antibodies have been provided second hand e.g. *via* colostrum or *via* an injection of serum as in lamb dysentery.

Keel mark: the marking of sheep with a coloured material. It usually refers to the marking of the sternum of rams to indicate whether or not they are working.

Kemp: a coarse hair-like fibre which does not take dye readily.

Marginal: an economic term applied to land which is near the verge of profitable agricultural use under a particular system.

Mastitis: inflammation of the udder.

Metabolisable energy (M.E.): that part of the energy ration the animal can use to do useful work, i.e. keep warm, grow wool, produce milk.

Metritis: inflammation of the womb.

Oestrus: the time of heat in the female when she will accept service from the male.

Ovulation: the shedding of eggs from the ovary.

Parturition: the act of giving birth.

Perinatal: around and near the time of birth.

Ram: an entire male sheep.

Repeatability: the degree to which an animal repeats a trait during its lifetime, e.g. litter size, milk production.

Rhizobium: a genus of nitrogen-fixing bacteria which live symbiotically within the roots of leguminous plants.

Rig: a ram with an undescended testicle.

Rise: (in wool) the time when the wool starts to grow and the yolk rises in the fleece.

Rumen degradable protein (R.D.P.): protein which is broken down in the rumen to much simpler substances.

Rumen undegradable protein (U.D.P.): protein which passes unchanged through the rumen.

Serum: a preparation from the blood of an animal used to provide disease antibodies for another animal. When infected a serum gives immediate passive immunity to the recipient.

Shearling: a sheep which has undergone one shearing at the normal time.

Shedder: a system of gates in a race whereby sheep can be separated into specific groups.

Sheep sick: land so heavily infected with sheep parasites and disease organisms as to be dangerous to sheep.

Shieling system: a historic method of exploiting grazings in the Scottish highlands whereby cattle and sheep were moved on to the mountains for the summer to return to the lowground settlement for the winter.

Steaming up: the term applied to the extra feeding of a female mammal immediately prior to parturition in order to stimulate milk production.

Stint: a limit applied to the number of animals a rights holder may run on a common grazing.

Stocking density: the number of sheep carried per unit area of a particular field or grazing at a particular time.

Stocking rate: the total number of sheep divided by the total area of pasture devoted to their support over a longish period of time, usually a year.

Store: a young animal intended for meat and requiring further feeding.

Stratification: the layering effect caused by the movement of sheep and their progeny from the hills to the low ground.

Suint: dried sweat of the sheep.

Symbiosis: the living together of two specifically different organisms to their mutual advantage.

Tup: an entire male, sheep, i.e. a ram.

Tupping: the mating of a female sheep by a male.

Two-toothed: a sheep with two permanent incisors, approximately eighteen months of age, i.e. a shearling.

Vaccination: the act of inoculating an animal with a killed or attenuated culture of disease organism to stimulate antibody formation.

Vasectomised: a vasectomised ram is one in which the vas deferens from each testicle has been tied off in a surgical operation rendering the ram infertile but unimpaired in his ability to serve a ewe.

Wether or wedder: a castrated male sheep.

Yeld or eild: a barren ewe.

Yolk: a combination of suint and grease found in sheep's fleece.